A FIELD GUIDE TO URBAN PLANTS

The following symbols are used as a quick guide

⌘	Four petals or fewer	⌘	Woody plant
⌘	Five petals	⌘	Grass or rush
⌘	More than five petals	⌘	Fern
⌘	Mirror-symmetrical (zygomorphic) flowers	⌘	Moss or liverwort

The colour of the above symbols
reflects the flower colour in each case.

ALEXANDRA-MARIA KLEIN
JULIA KROHMER

A FIELD GUIDE TO URBAN PLANTS

COMMON SPECIES OF PAVEMENTS, WALLS AND WASTE GROUND

PELAGIC

CONTENTS

FOREWORD: GREAT VARIETY ON A SMALL SCALE

Biodiversity is an almost infinitely broad field. Most people instinctively think of distant rainforests or coral reefs, African savannas or lush mangrove swamps. Yet relatively few appreciate the diversity and exquisite beauty of our natural world here at home, indeed right outside our front door – in the cracks between paving stones and the joins among the kerbstones of our inner-city areas.

A mapping of town habitats carried out by a Master's student on behalf of the Senckenberg Society for Nature Research in 2014 recorded over 300 species. In Germany as a whole, the total would number over 500. This flora, so often ignored or simply regarded as a mass of weeds, is most certainly worth a second look. One can only wonder at the incredible tenacity and adaptability of such species, and at the same time their often delicate beauty. They provide a habitat for many small creatures and have any number of positive side-effects for infrastructure, climate and human society.

To get to know and enjoy this diversity is to appreciate its value to our fellow citizens. It can and should be a first step on the way to a deeper interaction with our flora and fauna, so urgently needed today, literally right outside our front door and accessible to one and all. The charming accessibility of awareness campaign such as #morethanweeds is a wonderful way of bringing this variety to a wider audience in a playful and creative style.

I am therefore delighted by the publication of this book, an attractive introduction to this topic with the one hundred most important species to be found on our pavements and in our squares. I am confident it will find a wide and appreciative readership.

Professor Katrin Böhning-Gaese

Scientific Director, Helmholtz Centre for Environmental Research GmbH – UFZ, Leipzig

PREFACE: MOSS GARDENS ON YOUR DOORSTEP

The Kokedera (Moss Temple) in Kyoto is famous throughout the world. The Japanese word 'koke' does not mean moss in the botanical sense, but covers any small, delicate plants such as many of those that can be found in our streets and on walls.

My research has always involved mosses, many of which thrive in places typical of urban plants: cracks between paving stones, in walls and in other urban habitats that we too easily overlook. But once you have immersed yourself in the fascinating microcosm of these plants, you will start to see them everywhere. It is precisely such an awareness that this book wishes to promote by sharpening our perception of these inconspicuous and yet vital plants.

At the University of Freiburg, we are hoping to introduce flower areas around the buildings that are dotted across the town and anticipate the discovery of many exciting species. A further bonus will be that the university buildings will then be linked by a network of pollinator corridors.

I am sure this book will make a significant contribution to a greater awareness of species and hope that it will stimulate people to open their eyes to the natural world.

Professor Stefan A. Rensing

Vice-chancellor for Research and Innovation, Albert Ludwig University Freiburg

INTRODUCTION

BUILDING AWARENESS

———

I beheld a "weed"
But when its name was revealed
True beauty I saw
(Japanese Haiku, loosely translated)

BLINDNESS TO PLANTS

Most people can recall the last wild animal they saw. But is that the case with a plant outside the house? Or perhaps a tree by the side of the road? A hedge? A particular flower? If we notice plants at all, mostly it is just as a pleasant backdrop or as urban greenery. Very few people take a closer look, spending a moment examining the shape of the leaves or the interesting features of the flower. Are there any insects visiting the plant?

At the end of the twentieth century, the American botanists Elisabeth Schussler and James Wandersee coined the phrase 'plant blindness', by which they mean 'the inability to see plants in their own environment'. This is not only regrettable but also dangerous:

it implies a failure to appreciate the value of plants and therefore a lack of interest in their protection and conservation. This is especially serious in towns, where we desperately need plants to alleviate the effects of pollution and climate change. Indeed, we need plants, both big and small,

What plants have you noticed recently in your neighbourhood?

Miniature ecosystems soon develop in the cracks between paving stones.

not only trees and hedges in parks, but also the innumerable small grasses, mosses and other plants that inhabit cracks, gaps and corners.

VALUABLE ECOSYSTEMS

Our towns and inner cities are dominated by concrete and tarmac, but if you look more closely you will find plants almost everywhere. Right under your very feet! Often it is the rubbish, broken glass, chewing gum and innumerable cigarette ends that catch the eye. Yet in amongst all this you will discover small, stubborn, stoical fragments of green. Between the paving stones, in the cracks of gutters and in the crevices of walls these little botanarchists have dug themselves in. A multitude of herbaceous plants, grasses, bryophytes, ferns and seedlings of trees and shrubs have found a niche, adapting to the extreme conditions. In order to do so, they must defy pressure from pedestrians, traffic, drought, heat, the sun's radiation, soil compression, salt spreading and other pollution, and not least the street-cleaning vehicles and all too often an obsession with 'tidiness'. Despite all this, many survive to provide microhabitats for numerous insects and other organisms.

KERB SQUATTERS – BEAUTIFUL AND USEFUL

These interlopers have many uses to us humans: a thick growth in the cracks between the paving stones increases their stability; green crevices absorb surface moisture, help the water to seep away, contribute to cooling and bind in the soil and dust. In addition, wild plants in our towns play a significant role in the urban ecosystem, offering food and protection to other organisms such as wild bees, beetles and ants. Moreover, a second look (if one is needed) will reveal their beauty, not only of their petals but also the shapes of their leaves, and how different species grow together.

Intriguing assemblages of species can arise in urban micro-niches. Here three species of fern coexist in the same crevice in an old wall: Rustyback, Maidenhair Spleenwort and Hart's-tongue.

WANDERING WITH PURPOSE

Casually observing the plants growing in your town is a simple and enjoyable way to learn more about nature, but if you wish, there are various other things you can do to extend your knowledge and involvement. For instance, look out for the awareness-raising campaign #morethanweeds, which is counterpart to the German #krautschau [look plant!]. Another fun thing to engage with is #wildflowerhour. Local-interest societies and websites (such as https://wildbristol.uk/) are another great way to discover what's out there. For those enthused enough to take their interest in plants even further, the Botanical Society of Britain and Ireland offers a wealth of resources, courses, expertise and enthusiasm that help to inspire and inform.

Finally, do remember this this book is only an introduction. Almost any species could appear on a street near you, and this one of the main delights of urban botany. So keep looking... and you might come across something that no one else has seen in before in this ever-changing and always fascinating environment. Happy hunting!

DIVERSITY IN TOWN

Towns invariably conjure up an image of concrete, tarmac and glass. But a closer look will reveal an extraordinarily rich variety of plants and animals.

Why are urban places so unexpectedly biodiverse diverse? In the perception of most people, towns and wildlife are at opposite ends of the spectrum. Urban areas, after all, are largely made up of artificial structures and building materials, a fact that is hard to escape. However, towns are also endowed with surprisingly rich biodiversity – as is shown, for instance, by the over 500 species of wild plants being found in the cracks of German pavements. One reason for this is that the sheer range of structures and land-uses offers innumerable habitats to plants and animals. Verges, flowerbeds, street trees, green spaces, hedges and old walls, green rooves, undeveloped building plots, gardens, cemeteries, parks and more all make vital contributions to urban habitats. Even the buildings themselves accommodate many species, such as birds and bats. In fact, towns are nowadays more species-rich than many areas in the countryside because, compared with intensive and industrial modern farms, they offer a wide variety of niches attractive to wildlife. Towns and cities, it turns out, are fast becoming a refuge for wild species sadly no longer welcome in much of the wider landscape.

Façades covered with vegetation provide a home for a whole variety of wildlife.

Our roads and railways offer corridors for plants to spread.

For thousands of years, towns have been commercial centres and thus key meeting places for people from far and wide. At these crossroads of humanity, seeds from all over the world are able to establish themselves and spread. Garden plants colonise new locations and, when their novelty has worn off, exotic animals are released by their weary owners – it is especially the heat-loving species that are able to take advantage of the warmer urban climate, also known as the urban heat island effect. This is why towns all over the world have a large proportion of non-native plants and animals, and also the reason we will encounter many so-called neophytes, or non-native species, in this book.

VEGETABLES AND CEREALS

Even tomato, wheat, millet and sunflower seedlings can often be found on street corners. It may have been that someone threw away their kebab with tomatoes or was feeding seeds to the birds. How good it would be if these seedlings could grow to full size and provide us with fresh food in the town! Of course, they need soil to flourish, whereas the gaps in the urban environment are usually too small and dry and are compacted by being driven over and walked on

A bumblebee on a tomato flower – plants in the town may provide nourishment for both animals and humans.

continuously. It might be feasible, though, to exploit the areas around our urban trees to grow our food – as can be seen in occasional examples of guerilla gardening.

In the face of climate change, some of our food plants will do better under trees than in bright sunlight. Urban vegetable gardening in raised beds across a network of small public areas throughout our towns also adds to biodiversity. At the same time, we town dwellers can become reacquainted with how our food is produced.

The fundamental message here is that all plants are useful to something and that each species has its own beauty and wonder just waiting to be properly appreciated.

Flower strips with a variety of appropriate native species should be a standard feature in our towns. Careful management of verges can also allow a good range of species to reappear from the soil seed bank.

LONG SLENDER PODS

VERY SMALL CREAMY-WHITE FLOWERS

ROSETTE WITH SPOON-SHAPED LEAVES

Adapted to nutrient-poor habitats and does very well in towns. It is usually self-pollinating and seeds are spread by soil movement, wind or their adhesive threads.

THALE-CRESS
Superstar of plant science

Arabidopsis thaliana | Brassicaceae

This inconspicuous species is surely the world's best-researched plant. It has been a model organism in genetics since 1940 and, thanks to its simplicity, was the first plant genome to be sequenced. This species is native to the temperate zones of Eurasia and North Africa, but it can now be found across large parts of the world. Originally, its natural habitats were dry woodland clearings, but it took advantage of the beginnings of human agriculture and has flourished ever since. Easy to miss, it is surprisingly beautiful when examined up close.

DESCRIPTION: upright, mostly branching, decreasingly hairy towards top of stem | **GROWTH FORM:** annual or biennial | **HEIGHT:** 5–30 cm | **LEAVES:** rosette leaves longish spoon-shaped, serrate with a blunt tip, stem leaves mostly entire, without petioles, narrowly lanceolate; the whole plant has a blue-grey hue | **FLOWERING:** ✤ Apr–May | **FRUIT:** long slender pods, rounded in cross-section

CLUSTERS OF WHITE FLOWERS AT THE APEX OF THE STEM

LEAVES SMELL OF GARLIC WHEN CRUSHED

SQUARE STEMS

GARLIC-MUSTARD
Flavoursome travel companion

Alliaria petiolata | Brassicaceae

Often known by the name Jack-by-the-Hedge, this attractive, fresh-green plant grows across large parts of Europe, the Middle East and Central Asia as far as China and India. It was introduced to North and South America as an aromatic and medicinal plant and is now categorised as an invasive species there. It grows in spring in cool, nutrient-rich places – particularly in open deciduous woodland, but also in towns, parks and along path edges. Often occurs together with Stinging Nettle *Urtica*

DESCRIPTION: fresh-green plant with upright, square stem and clusters of small white flowers at the apex | **GROWTH FORM:** herbaceous biennial | **HEIGHT:** 20–100 cm | **LEAVES:** kidney-shaped, notched basal leaves with long petioles; alternate, heart-shaped stem leaves; toothed margins | **FLOWERING:** Apr–Jun | **FRUIT:** pods

dioica which also has a preference for nitrogen-rich habitats.

Garlic-mustard provides food for many different species of insect.

PREFERABLY EATEN FRESH AND RAW, since the characteristic peppery garlic-like taste disappears with cooking or drying, unlike the better-known Wild Garlic *Allium ursinum* – a species of woodlands and shady banks. Nonetheless, this plant has been used for seasoning food for about 6,000 years. We know this from the finds of plant remains on pottery sherds in northern Germany and Denmark. Garlic-mustard is thus the oldest known European condiment. In the Middle Ages, it was often grown in the garden because real pepper imported from Asia was too expensive for most people. Young leaves taste good in smoothies, soft cheese or simply eaten with a slice of bread. The still-green pods can be added to create a peppery salad, whilst the ripe black seeds may be used like pepper or made into a kind of mustard. The pungent root may be grated like horseradish and the flowers are excellent as salad decoration. Everything about this plant is delicious!

ALSO LOVED BY INSECTS

The caterpillars of several species of butterfly feed on the leaves of Garlic-mustard, notably the Orange-tip – that wonderful symbol of spring. Many wild bees, beetles, flies and hoverflies like to help themselves to the easily accessible nectar that collects at the base of the flower. However, this species is not so friendly to other plants: its roots give off chemical substances that inhibit the growth of mycorrhizal fungi. The seedlings of many woody plants depend on the latter, and without them grow poorly or not at all. Garlic-mustard thus prevents woody plants from growing too close to it.

The seeds are distributed by the wind. In rainy weather they become slimy and sticky and get stuck to passing walkers or animals, hitching a ride to a new location.

ROUNDED PETALS WITH NO NOTCH

DENSE CLUSTERS OF FLOWERS

STALKLESS LEAVES

The hairs protect the plant from sunlight and evaporation. It has flat hairs visible with a good magnifying glass.

SWEET ALISON
Garlands among the paving slabs

Lobularia maritima | Brassicaceae

Native to the Mediterranean and North Africa where it favours coastal cliffs, this popular garden plant has been grown in Britain since at least the early 1700s. Familiar as an escape in seaside areas, towns and cities, on pavements and sometimes below hanging baskets. The flower size and colour are quite variable due to all the different cultivated varieties, but once naturalised it tends to revert closer to the wild form. The flowers have a sweet, pleasant smell – hence the name. 'Alison' is in fact a bastardisation of *Alyssum*, the genus in which this species used to be placed.

DESCRIPTION: upright or decumbent, branching below; slender grey-green leaves densely covered in short hairs | **GROWTH FORM:** annual or perennial | **HEIGHT:** 10–35 cm | **LEAVES:** sessile or with a broad petiole, linear-oblanceolate up to 5 cm long, entire, sometimes fleshy, tip obtuse | **FLOWERING:** ✿ Apr–Oct | **FRUIT:** small flattened oval pods

SMALL HEART-SHAPED PODS

SMALL WHITE FLOWERS

STEM LEAVES LANCEOLATE AND SESSILE

Recipe for success: one plant produces up to 64,000 seeds which remain viable for up to 30 years. Four generations can be created each year.

SHEPHERD'S-PURSE
Chic purse for the town dweller

Capsella bursa-pastoris | Brassicaceae

This ubiquitous and familiar species flowers almost through the whole year. It likes to grow in bright and fertile situations, and owes its name to the shape of its seedpods which are reminiscent of the small pouches carried by medieval shepherds. They taste hot and spicy and were once used like pepper to season food. A sticky slime around the seeds aids their distribution as they attach themselves to the soles of shoes and the fur and paws of animals. This slime also contains special enzymes which break down nematodes near the germinating seeds, protecting them while at the same time creating additional nutrients in poor soils.

DESCRIPTION: inflorescences initially somewhat clustered, but as the fruits develop the flowerheads become longer, loose racemes | **GROWTH FORM:** annual or biennial | **HEIGHT:** 2–70 cm | **LEAVES:** basal leaves with short petioles, undivided to pinnatifid, unevenly serrate (reminiscent of a dandelion), stem leaves lanceolate, sessile | **FLOWERING:** ✿ Jan–Dec | **FRUIT:** small heart-shaped pods

USUALLY JUST FOUR STAMENS

LONG THIN PODS

CONVEX ROUNDED LEAF ROSETTE

PINNATE LEAVES

Thanks to a special mechanism, this pretty little plant can launch its seeds over distances up to 1.4m. In the colder months it will readily self-pollinate, too.

HAIRY BITTERCRESS
Urban salad vibes

Cardamine hirsuta | Brassicaceae

The natural distribution of this formerly quite rare, extremely adaptable species is Europe and Asia, but today it can be found almost worldwide and is equally at home in vegetable patches, in the cracks between paving stones, on building sites or in flowerpots. Its roots may be up to 35 cm deep and new leaf rosettes can be found very early in the year (indeed, they can appear at any time really). Good for us, because its tender leaves and shoots are edible and rather flavoursome – they taste aromatically pungent, similar to its relative, Garden Cress *Lepidium sativum*, and can be used in the same way.

DESCRIPTION: basal rosette that appears convex. Several branching stems | **GROWTH FORM:** annual | **HEIGHT:** 7–30 cm | **LEAVES:** basal and stem leaves with 1 to 7 pairs of pinnate leaves and a terminal leaflet, pinnate leaves of the basal rosette kidney-shaped or widely obovate, those of the stem leaves elongate ovate | **FLOWERING:** Mar–Jun | **FRUIT:** slim cylindrical pods

INTENSE FOETID-PEPPERY ODOUR

PAIRED, ROUNDED, ROUGH PODS

ATTRACTIVE PINNATE LEAVES

LESSER SWINE-CRESS
A real stinker?

Lepidium didymum | Brassicaceae

The origins of this species are unknown but thought to be South America. It has long been a common wayside and waste ground plant in Europe and else-where, and was first recorded in Britain in the early 1800s. A mainstay of damp farm gateways, it is also frequent in towns and cities, in pavements, paths and gardens. Often possible to smell it before you see it: the pungent peppery odour is unique – and although most people find it unpleasant, some are strangely fond of its particular scent. When seen in a mass, its intricate leaves have a certain beauty. **Swine-cress** *Lepidium coronopus* is similar but generally hairless, darker green in colour and has distinctive warty fruits.

DESCRIPTION: rosettes of pinnate leaves, usually with at least sparse hairs, crowded racemes of minute whitish flowers | **GROWTH FORM:** annual | **HEIGHT:** procumbent or ascending, up to 40 cm | **LEAVES:** basal and stem leaves with 2 to 6 pairs of pinnate leaves and a terminal leaflet, leaflets often acutely lobed anteriorly | **FLOWERING:** May-Oct | **FRUIT:** pairs of tiny rounded pale green pods

THICK SUCCULENT LEAVES

PETALS WIDELY SPACED

PINKY-WHITE FLOWERS

DANISH SCURVYGRASS
Roadside Milky Way

Cochlearia danica | Brassicaceae

In recent years, this native species of sandy and pebbly shores has made dramatic advances inland along our road systems. Its ecological status as a halophyte (that is, being salt-tolerant) has allowed it to exploit road-edge habitats where salt-grit is spread in the winter – along with other species formerly confined to the coast, such as Reflexed Saltmarsh-grass *Puccinellia distans* and Sea Barley *Hordeum marinum*. Its clusters of lilac-white flowers make a striking impact in winter and early spring; by the middle of the year all that is left are empty seedpods and dead dried stems. The scurvy-grasses got their name from being used to ward off scurvy due to their high vitamin C content.

DESCRIPTION: stems ascending or spreading, sometimes branched, hairless | **GROWTH FORM:** annual to biennial | **HEIGHT:** 3–25 cm | **LEAVES:** rounded and heart-shaped at base, bluntly lobed above and nearly sessile higher up the stem, dark green or purplish, fleshy | **FLOWERING:** ☘ Dec–May | **FRUIT:** small rounded oval pods

THE PETALS ARE DEEPLY DIVIDED – DESPITE APPEARANCES THERE ARE ONLY FOUR, NOT EIGHT.

LEAVES CONFINED TO BASAL ROSETTE

COMMON WHITLOWGRASS
Modest and miniature

Erophila verna | Brassicaceae

This dainty little plant likes a lot of light and grows in very nutrient-poor locations that are difficult for many species. It is so named because it was traditionally supposed to cure whitlows, infections of the fingertip. The seeds germinate in winter and the plant survives as a small rosette of leaves until early spring, when it produces small, white flowers. The leaves die back before the seeds are ripe – but the chlorophyll in the pods is sufficient to nourish the seeds until they are ready.

Whitlowgrass is the one of smallest species to be found in urban areas, perhaps also the most fleeting. It flowers very early in spring and fruits after only a few days, often disappearing again by April. For this reason it is one of a suite of species known as 'early-spring ephemerals'.

DESCRIPTION: basal rosette with unbranched, leafless stems with stellate hairs, often grows in clumps of several of the same species | **GROWTH FORM:** winter annual | **HEIGHT:** 3–15 cm | **LEAVES:** entire or slightly toothed, obovate to lanceolate | **FLOWERING:** Feb–May | **FRUIT:** small oval pods

FIVE ROUNDED PETALS

REDDISH STEMS WITH LOTS OF STICKY HAIRS

LEAVES WITH THREE LOBES

The name 'saxifrage', which is applied to numerous plants of different genera, means 'stone-breaker' and relates to the habit of growing out of cracks in rock.

RUE-LEAVED SAXIFRAGE
Footsteps of stars

Saxifraga tridactylites | Saxifrageaceae

A simply exquisite little native plant of rocky or sandy calcareous substrates. Like the previous species, it has also moved into urban areas and can be found constellating pavements and walls early in the year. The stems are a characteristic red colour and are covered in sticky hairs (don't be surprised to find dog hair tangled up in patches of this plant). The attractively lobed yellowy-green, red-tinged leaves resemble those of the medicinal plant Rue *Ruta graveolens*, hence the vernacular name (but the species epithet *tridactylites* – 'three-fingered' is helpful to remember).

DESCRIPTION: red zigzagging stems and thick-looking lobed leaves give this plant a distinctive appearance | **GROWTH FORM:** winter annual | **HEIGHT:** 3–15 cm | **LEAVES:** pinnately divided into 1 to 5 lobes | **FLOWERING:** ✿ Feb–May | **FRUIT:** capsule

SEPALS LONGER THAN PETALS

PLANT APPEARS GREY-GREEN

THYME-LEAVED SANDWORT
Scruffy asterisms

If you compare them to carnations in a flower shop, you might be a little disappointed, but take a closer look: on a sunny day the tiny white flowers glow like stars.

Arenaria serpyllifolia | Caryophyllaceae

Originally a native of the temperate zones of Eurasia and north Africa, this little herb can today be found over large areas of the world, its propagation being aided by human factors. In urban areas it favours walls, the edges of paths and cracks in paved surfaces.

While appearing to be fragile, it is in fact very resistant to trampling. Despite the passing resemblance of its leaves, this species is not related to thyme. The flowers can be self-pollinating or may be pollinated by small solitary bees.

DESCRIPTION: delicate, spindly, grey-green and rough-haired, branched growth with pretty white flowers | **GROWTH FORM:** annual or biennial | **HEIGHT:** 3–30 cm | **LEAVES:** opposite; oval or ovate with pointed tips, 2–6 mm in length | **FLOWERING:** May–Sep | **FRUIT:** capsule

MINUTE FLOWERS WITH VERY SHORT PETALS

OPPOSITE LEAVES

ROUNDISH CAPSULE FRUITS

PROCUMBENT PEARLWORT

An optical illusion?

Despite its humble appearance, Pearlwort was replete with folkloric significance: it could ward off evil, protect cattle and even keep fairies away from unborn children.

Sagina procumbens | Caryophyllaceae

At first glance it may appear to be a moss, but in fact this very small plant belongs, perhaps rather astonishingly, to the carnation family. It is an evergreen, cushion-forming species with tiny, star-shaped flowers. This hardy little plant is second to none in decorating the cracks in pavements and is abundant all over the British Isles. Not even a rough-and-tumble game of football played by children in front of the garage door will disturb this small, trample-resistant species. Pearlwort is very tolerant of salt, so the plant is not affected by its use on pavements and roads in the winter.

DESCRIPTION: small cushions or clumps reminiscent of moss | **GROWTH FORM:** groundcover perennial | **HEIGHT:** 2–15 cm | **LEAVES:** opposite, linear | **FLOWERING:** ❀ Jun–Sep | **FRUIT:** roundish capsule with four serrations

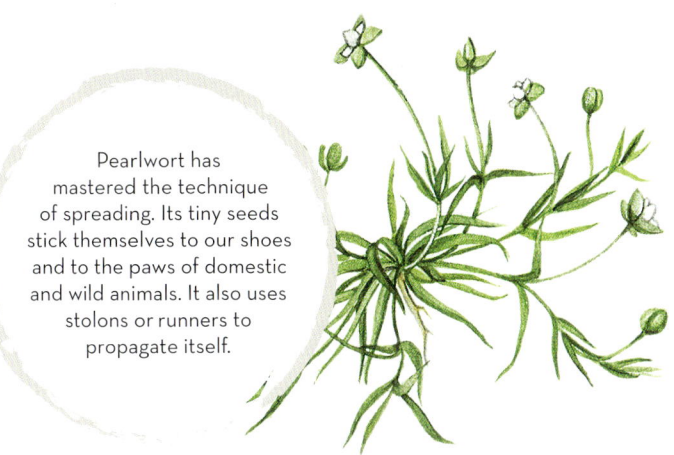

Pearlwort has mastered the technique of spreading. Its tiny seeds stick themselves to our shoes and to the paws of domestic and wild animals. It also uses stolons or runners to propagate itself.

WEEDS, SEEDS, SIDE SALAD

Procumbent Pearlwort spreads amongst the cracks between the paving stones and so is branded as a weed. But it bothers nobody; indeed, it's so small that most people do not even notice it. Its very attractive cousin Heath Pearlwort *Sagina subulata*, with its beautiful white decorative starlike flowers, has succeeded in breaking into the horticulture industry, where it is marketed as a robust evergreen groundcover plant. Both species are also edible, should you be that way inclined, or just really really hungry.

The minute star-shaped flowers only open when the sun is shining.

CRINKLY EDGES TO FOLIAGE

FLOWERS IN UPRIGHT SPIKES

ROUNDED FLESHY LEAVES

NAVELWORT
Green money from a hole in the wall

Umbilicus rupestris | Crassulaceae

A favourite with foragers, the fleshy leaves of Navelwort – or Wall Pennywort, as it's often known – have a pleasant, slightly salty, cucumbery flavour. Common in the cooler, damper environs of the west but rare in the east, this member of the stonecrop family loves shady walls and stonework. With its long spikes of many tubular flowers and frequently bearing a reddish hue, it is a striking plant. The catalogue of local and folk names is long, including Batchelor's Buttons, Cups-and-Saucers, Dimple-wort, Lover's Kinks, Lucky Moon, and Pancakes.

DESCRIPTION: hairless, succulent, whole plant often tinged reddish | **GROWTH FORM:** perennial | **HEIGHT:** up to 40 cm, sometimes more | **LEAVES:** rounded, with a slightly crinkly edge, stem goes to the middle of the back of the leaf | **FLOWERING:** ✿ crowds of greenish-white – sometimes pinkish-brown – flowers with five fused petals (like little tubes) on a long upright spike, Jun–Sep | **FRUIT:** chestnut brown, splitting, falling from dry flower tube

CYLINDRICAL LEAVES

STAR-SHAPED WHITE FLOWERS

OFTEN TINGED RED

Look closely at the plant stems and you may be able to spot the roots. This kind of vegetative propagation frequently occurs in sedums.

WHITE STONECROP
A born survivor and pioneer

Sedum album | Crassulaceae

Colonises cracks in walls and stony waste places. The leaves store water and are coated with a layer of wax that reduces evaporation – a perfect adaptation to strong sunlight and exposed conditions. Thanks to its shallow roots, the plant also thrives in crevices in walls. Butterflies such as Small Tortoiseshell, Comma and Red Admiral are attracted to its flowers. The Hawthorn Mining Bee *Andrena chrysosceles* collects pollen from this species to feed its developing offspring.

DESCRIPTION: succulent with, white flowers with reddish anthers | **GROWTH FORM:** herbaceous perennial | **HEIGHT:** 2–20 cm | **LEAVES:** alternate, succulent, linear to cylindrical, green or reddish | **FLOWERING:** ❀ cluster with 2 to 6 flowers, May–Sep | **FRUIT:** follicle

DOWNWARD-POINTING BRACTS BELOW UMBELS

MUCH-DIVIDED, FEATHERY LEAVES

BRANCHING STEMS

FOOL'S PARSLEY
Look for the dangling needles

Aethusa cynapium | Apiaceae

As with many urban plants, this is another species that formerly was better known as a weed of cultivated ground. In his famous *Herball*, John Gerard (1545–1612) memorably described it as having a 'naughtie smell'. You would indeed be foolish to treat this like parsley, because the whole plant is poisonous – albeit not as toxic as some of our other umbellifers, such as **Hemlock *Conium maculatum***, which also appears in urban waste ground. The latter has tall, purple-blotched stems. Fortunately, neither species looks much like the beloved culinary herb.

DESCRIPTION: dark green, hairless leaves and stems, long upper bracts characteristically point downwards below the flowerheads | **GROWTH FORM:** biennial | **HEIGHT:** up to 60 m | **LEAVES:** alternate, bipinnate or tripinnate, feathery | **FLOWERING:** ✿ compound umbels 2-3 cm across, Jun–Aug | **FRUIT:** only slightly longer than wide, prominent thick ridges

WHITE COMPOUND UMBELS

THREE-CORNERED, SMOOTH AND HAIRLESS STEM

The leaves near to the ground survive mild winters – a competitive advantage in spring. The powerful runners force their way through almost anything and allow this plant to grow even in narrow cracks and crevices.

GROUND-ELDER
Troublesome, tasty and full of vitamins

Aegopodium podagraria | Apiaceae

A nuisance in the garden, Ground-elder has a reputation as an annoying weed that is really tricky to get rid of. However, it should rather be seen as a tasty source of vitamins and minerals and a plant loved by insects. In times of famine in Europe, Ground-elder was a lifesaver, while today it is valued by some as a wild vegetable. It is easy to see a whole variety of different insect species busying themselves on the flowers – this species provides a highly valuable habitat and food source.

DESCRIPTION: Prolifically growing rhizome and angular bare stem | **GROWTH FORM:** herbaceous perennial | **HEIGHT:** 50-90 cm | **LEAVES:** alternate and serrate, often pinnatifid on one side and undivided on the other | **FLOWERING:** compound umbels without bracts, Jun–Jul | **FRUIT:** schizocarp with two stigmas protruding from the top like antennae

STAR-SHAPED INFLORESCENCES

WHITE UMBELS

SMELLS OF CARROT WHEN CRUSHED

WILD CARROT
Mimics insects, clouds and stars

Daucus carota | Apiaceae

This species impersonates insects in order to attract other insects. The purple-black central dot consists of a dense group of flowers containing anthocyanin (dark plant dyes) in the middle of the umbel, which itself is made up of many individual umbellets (smaller clusters of flowers). Potential pollinators are attracted by the presence of the fake insect because they believe that there must be a good supply of pollen and nectar. With this trick, Wild Carrot is able to drawn in numerous insects ranging from flies and beetles to wasps and bees. This mock insect, however, is not found on all the flowers, so don't let that confuse you when you are trying to identify it. Star-shaped bracts are always present below the umbel. If you

DESCRIPTION: Dark flowers in the middle of the umbel, stellate bracts, upright growing, roughly hairy | **GROWTH FORM:** biennial | **HEIGHT:** 20–120 cm | **LEAVES:** alternate, bipinnate or tripinnate | **FLOWERING:** ❀ compound umbels, May–Jul | **FRUIT:** prickly schizocarp consisting of two parts

are still unsure, crush a leaf between your fingers. It will smell of parsnip or carrot.

UNTIDY BIRD'S NEST
Both before and after flowering, the umbel of the wild carrot looks like a bird's nest because the bracts curl up when wet. You can also recognise its domesticated cousin, the cultivated carrot, using this feature. In dry conditions, the umbels then re-open. This phenomenon is known as hygroscopic movement, that is, movement controlled by humidity.

EDIBLE, OF COURSE
The white roots, up to 80 cm in length, are quite woody in the second year and sometimes very spicy in taste, so it is best to eat them in the first year after germination. We can also recommend the sweet-flavoured young leaves and stems in salads.

The fruiting head looks a little like a bird's nest.

A characteristic feature is the single dark flower in the middle of the umbel that mimics a visiting insect.

Not only does the wind distribute the seeds, but they also attach themselves to squirrels and rats, dogs and cats and in this way are spread far and wide.

INCONSPICUOUS
FLOWERS

TRANSLUCENT
SKIN-LIKE LEAF
SHEATHS

CREEPING
RUNNERS

COMMON KNOTGRASS
Lies down to sleep

Polygonum aviculare | Polygonaceae

Knotgrass is extremely resistant to trampling and at night folds down its upright leaves to protect them against nocturnal troublemakers. This phenomenon is called nictinasty. The small, inconspicuous, white-pink flowers contain no nectar. If you put bags over the flowers to prevent insects from pollinating them, it can be observed that many seeds still develop. This strongly suggests that self-pollination is occurring, which is in keeping with

DESCRIPTION: Leaf sheath skin-like, translucent, often shiny silver-coloured, 1 to 3 flowers, prostrate to upright | **GROWTH FORM:** biennial | **HEIGHT:** 5–50 cm | **LEAVES:** alternate, undivided, sessile or with short petioles | **FLOWERING:** ✿ May–Nov | **FRUIT:** red, oval nutlet, which may remain viable in the seedbank for many hundreds of years

the fact that the flowers are unattractive and unassuming. Common Knotgrass usually nestles down flat in the smallest nooks and crannies on heavily used footpaths, but in more favourable locations with better soil it can also develop into substantial plants up to 50 cm in height. It can withstand very dry conditions everywhere.

Common Knotgrass creeps along the ground and can form messy, straggling mats.

BACK IN THE MIDDLE AGES

German polymath Hildegard von Bingen listed knotweed as a medicinal plant. It contains substances such as silicic acid and flavonoids, in particular avicularin, as well as essential oils, tannins and resins. Even today, in traditional medicine the plant is still used internally for respiratory diseases and nosebleeds, and externally to help heal wounds.

NUTRITIONAL PLANT

Knotgrass was once used as bird and pig feed on farms throughout Central Europe up to an altitude of 2,200 metres above sea level.

Rainwater, sparrows and even scuffing footsteps spread the seeds across pavements, through paving cracks and along paths.

WHITE FLOWERS WITH YELLOW STAMENS

DARK-TINGED STEMS

This species is well adapted to many different habitats and in urban areas; it likes growing on roadsides, rubbish tips, railway cuttings and next to hedges or areas of water. However, it is not tolerant of long periods of drought.

BLACK NIGHTSHADE
Relative of the tomato, but with a dark name

Solanum nigrum | Solanaceae

Poisonous, pea-sized berries mean that this plant must be treated with care. The unripe fruits, in particular, contain considerable amounts of the alkaloid solanine. In other parts of the world, however, the ripe berries or even the leaves are eaten, and in many cultures the plant is also used as a medicinal herb. It is closely related to tomatoes and potatoes, as you can see from its pretty flowers. The geographical origin of Black Nightshade is not yet entirely clear, but it is thought to be from Eurasia.

DESCRIPTION: stem often tinged with blackish purple, graceful white flowers in loose groups | **GROWTH FORM:** annual | **HEIGHT:** 10–80 cm | **LEAVES:** striking dark green, ovate to lanceolate, leaf edges entire to curved serrate | **FLOWERING:** ✿ Jun–Oct | **FRUIT:** black berries (green when unripe)

LINE OF HAIRS ON STEM

FIVE DEEPLY CUT PETALS MAKES IT APPEAR AS THOUGH THERE ARE TEN

Chickweed has been spread around the world by humans since the Stone Age and has even followed us into town. It is a good indicator of nitrogen-rich soils.

COMMON CHICKWEED
A healthy weed

Stellaria media | Caryophyllaceae

The seeds of this familiar species ripen almost all year round and are able to germinate even at just a few degrees above zero. This enables it to produce two to three generations per year and quickly cover suitable locations with a dense carpet. In turn, this protects the soil from erosion and drying out. It is eaten by birds and is also healthy for us: it tastes good as a salad and contains a lot of vitamin C. Its ingredients have an anti-inflammatory, digestive, diuretic and detoxifying effect. Common Chickweed goes by many other names, among them chickenwort or winterweed.

DESCRIPTION: sprawling, branched, root-forming runners, one or two lines of hairs running down the stem | **GROWTH FORM:** annual | **HEIGHT:** 3–40 cm | **LEAVES:** broad ovate to pointed, small stems, unstemmed higher up, up to 1 cm long and 0.5 cm wide | **FLOWERING:** ✿ Jan–Dec | **FRUIT:** capsules

Photo labels: FIVE NOTCHED PETALS · OVAL-SHAPED LEAVES · WHOLE PLANT IS WELL HAIRY

COMMON MOUSE-EAR
Soft and friendly

Cerastium fontanum | Caryophyllaceae

A very variable and highly cosmopolitan native species of grassland, gardens, waysides and waste ground, its rounded leaves are easy to recognise. The whole plant is usually covered in dense long hairs, perhaps increasing its resemblance to a mouse. The five petals are each deeply divided into two lobes. Look out for **Sticky Mouse-ear *Cerastium glomeratum*** too, which has sticky glandular hairs and a more crowded inflorescence.

DESCRIPTION: mat-forming, tufted or straggling | **GROWTH FORM:** perennial | **HEIGHT:** to 40 cm | **LEAVES:** broadly obovate to pointed, grey-green, sessile | **FLOWERING:** ✿ Mar–Oct | **FRUIT:** capsules

WHITE TRUMPET-LIKE FLOWERS

SPEARHEAD-SHAPED LEAVES

Propagates through runners or seeds, which may be carried away in a river and drift intact for as much as nearly three years(!) before germinating in some far distant place.

HEDGE BINDWEED
Delightful ornamental or annoying thug?

Calystegia sepium | Convolvulaceae

This species is a good indicator of bad weather as the white flowers remain closed during rain and cold. At night they open, as they are pollinated mainly by nocturnal insects such as the aptly named Convolvulus Hawkmoth with its 8-cm-long proboscis. Then in the mornings you can often find various mining (*Halictus*) and sweat bees (*Lasioglossum*) busying themselves in the flowers. When climbing, the tendrils perform circular searching movements in an anti-clockwise direction, a single rotation taking just under two hours. Hedge Bindweed can live for several years. The resin in its underground tubers has a laxative effect. In some urban areas, the larger non-native **Large Bindweed *Calystegia silvatica*** (flowers up to 9 cm long) can be more common.

DESCRIPTION: twining plants with pure white trumpet-shaped flowers of fused petals up to 6 cm long | **GROWTH FORM:** climber | **HEIGHT:** 100–300 cm | **LEAVES:** heart-shaped or arrow-shaped, alternate, over 10 cm in length, on long supple stems | **FLOWERING:** ❀ Jun–Sep | **FRUIT:** capsules

SMALL COMPOSITE UMBEL-LIKE INFLORESCENCE

FINE PINNATE LEAVES

Yellow to grey seeds about a millimetre in size are distributed by the wind and are blown from one street corner to the next. Yarrow also propagates by means of runners.

YARROW
Appeals to insects and livestock

Achillea millefolium | Asteraceae

Some breeds of sheep love the leaves and flowers, tearing them off and leaving behind the tough stems that are not to their taste. As there are usually few sheep in urban areas, these attractive white-pink flowering plants are able to spread profusely! With plenty of pollen and nectar, Yarrow is favoured by flower-visiting insects, while numerous moth species rely on it as a foodplant for their larvae. Its soft feathery foliage is also used by birds to line their nests. The scientific name of this ancient medicinal plant goes back to Achilles, the hero of the Trojan Wars. He is said to have treated wounds with a tincture of Yarrow. In herbal medicine, it is employed mainly to treat gastrointestinal complaints. An old Somerset name was 'Goose-tongue', a reference to the spikes along the edges of the tongues of geese.

DESCRIPTION: a flat-topped cluster of flowers | **GROWTH FORM:** herbaceous perennial | **HEIGHT:** 20–120 cm | **LEAVES:** alternate with finely serrate margins, bipinnate or tripinnate | **FLOWERING:** ❀ Jun–Oct | **FRUIT:** achene containing a single seed

LONG INFLORESCENCE

SMALL FLOWERHEADS

LANCEOLATE LEAVES

The fluffy-haired seeds may be carried for long distances by the wind or attach themselves to the fur of pets and wild animals.

CANADIAN FLEABANE
Herbicide-resistant superweed

Erigeron (Conyza) canadensis | Asteraceae

Originating in North America, this is a rapidly spreading species that you are certain to find along pavement cracks and at the bases of walls. For insects, Canadian Fleabane is of little value and some conservationists pull it out of the ground – to little avail – wherever they see it because it displaces native plants. In the USA, it has developed glyphosate resistance and become a superweed. There are three other similar species to look out for, too: **Guernsey Fleabane** *Erigeron sumatrensis*, **Bilbao Fleabane** *E. floribundus* and **Argentine Fleabane** *E. bonarensis*, the last of which is much less common than the other species.

DESCRIPTION: very long panicle with many long, broad, closed involucres | **GROWTH FORM:** annual | **HEIGHT:** 20–100 cm | **LEAVES:** alternate lanceolate leaves | **FLOWERING:** ✿ Jul–Oct | **FRUIT:** achene with yellowish tuft of pappus hairs

LEAVES ARE WOOLLY

WHOLE PLANT IS CONSPICUOUSLY GREY-GREEN

THE URN-SHAPED FLOWERS ARE GOLDEN BROWN

JERSEY CUDWEED
Ultra creamy street crème-caramel

Laphangium luteoalbum | Asteraceae

Formerly a rare arable weed on the Channel Islands and in the southern counties, this species is swiftly becoming established in pavements of towns and cities, especially London. With its woolly appearance and cuddly flowers, it looks like no other plant found in this context. The reasons for the recent sudden and big increase in its distribution and abundance are unclear but could involve our warming climate, movement of soil for building purposes, accidental introductions with horticulture, or a combination of these factors. This plant and other cudweeds are eaten in parts of Asia.

DESCRIPTION: covered in thick woolly hairs, the whole plant is grey-green | **GROWTH FORM:** annual | **HEIGHT:** to 50 cm | **LEAVES:** alternate lanceolate furry leaves | **FLOWERING:** ✸ clusters of closed urn-shaped involucres, Apr–Jul | **FRUIT:** achene with pappus hairs

BUTTON-LIKE FLOWERS

OPPOSITE LEAVES

CLEARLY VISIBLE LEAF VEINS

The fruits have barbs on the calyx and attach themselves to animals or the clothes of passersby. Gallant-soldier is thus very successful in using burrs to disperse its seeds.

GALLANT-SOLDIER
Wanderer from the Andes

Galinsoga parviflora | Asteraceae

So say, this species marched with the French army under the leadership of Napoleon through Europe at the beginning of the nineteenth century. It probably simply escaped from a garden in Paris and spread across Europe at around that time. In English, the alternative name of 'quickweed' pays tribute to its ability to self-propagate at great speed. It was also cultivated for its tasty leaves that are packed with nutrients and minerals including iron, calcium, magnesium, zinc and vitamins A and C. The closely related **Shaggy-soldier** *Galinsoga quadriradiata* is actually more common in much of Britain and Ireland now; that species is hairier, its receptacle scales usually do not have lobes and the pappus hairs have a terminal projection.

DESCRIPTION: three-lobed ray florets widely spaced around the edges of the flower, dense central disc florets | **GROWTH FORM:** annual | **HEIGHT:** 10–60 cm | **LEAVES:** alternate, ovate and only hairy on the edge and on the veins of the underside | **FLOWERING:** May–Oct | **FRUIT:** achene

YELLOW DISC FLORETS, WHITE RAY FLORETS

LEAFLESS STEM

FLAT-LYING BASAL ROSETTE

DAISY
A universal favourite

Bellis perennis | Asteraceae

The most recognised plant in Europe is surely the Daisy, followed by the Dandelion. This is not surprising as Daisies grow almost everywhere and bloom all year round in warm regions. The flowers remain intact at temperatures as low as -15°C, turning pink after cold nights.

MANY FLOWERS IN ONE BASKET
To the layperson the Daisy appears to be only a single flower. But it is in fact a false flower, or pseudanthium, an inflorescence that consists of numerous individual flowerlets. This is the case with all species in the daisy family. The flowerhead moves as it orientates

DESCRIPTION: basal rosette with a leafless stem and a compound head consisting of many florets | **GROWTH FORM:** herbaceous perennial | **HEIGHT:** 5–15 cm | **LEAVES:** evergreen, simple leaves in basal rosette | **FLOWERING:** ✿ composite white or pink, Jan–Nov | **FRUIT:** achene

towards the sun and closes when there is rain.

Beetles, bees and flies visit Daisies, although insects mostly prefer other composite flowers if they are available nearby. However, Daisies can self-pollinate, so it is not a problem if insects seek nectar and pollen from other plants.

The leaf rosette lies flat on the ground, making it better able to survive being trampled.

The small seeds with wings are dispersed by wind and rain, as well as by earthworms, snails and people in towns. The daisy can also propagate vegetatively.

VERSATILE USES

The Daisy contains plenty of vitamin C, iron and magnesium as well as other less beneficial substances. Thus, it is always a good supplement to salads or soups. Daisies taste good on a piece of bread and butter or fried in some olive oil and added to pasta or salad. There is a whole variety of cookbooks with dozens of recipes. These flowers are also very attractive in wreaths and small posies. Daisies are furthermore used in many natural therapies, such as creams for reducing inflammation.

DURABLE The Daisy is very tough and can survive the wear and tear of well-trodden pavements and verges and so is often included in seed mixtures designed precisely for this type of environment.

CULTIVATED VARIETIES OF DAISY have been bred to produce larger and more generous flowerheads. Like tulips and roses, daisies have been grown to suit people's tastes. If insects could choose, they would of course go for varieties with more nectar and pollen!

Also look out for the garden escape **Mexican Fleabane** *Erigeron karvinskianus*, which is common on walls and among stonework especially in the south. It looks like a long, leggy Daisy, with more pink in the flowers.

GLOBULAR WHITE FLOWERHEADS

TYPICAL CLOVER LEAVES

Tolerates high nitrogen pollution, salt and being trampled on; strong creeping runners force their way along the cracks in the paving stones of streets and pathways.

WHITE CLOVER
Championed by bumblebees

Trifolium repens | Fabaceae

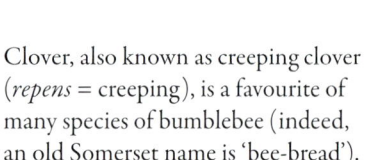

Extract of White Clover, with its phytohormone content, is reputed to alleviate the effects of the menopause. Indeed, Red Clover *Trifolium pratense* has an even higher level of this phytoestrogen that binds to the oestrogen receptors. However, little is yet known about the side-effects of phytohormones. Otherwise, White Clover, also known as creeping clover (*repens* = creeping), is a favourite of many species of bumblebee (indeed, an old Somerset name is 'bee-bread').

DESCRIPTION: creeping, with leaves that usually have pale chevron markings; globular flowerheads | **GROWTH FORM:** herbaceous perennial | **HEIGHT:** 15–45 cm | **LEAVES:** alternate, bare, elliptical to obovate finely serrate leaflets | **FLOWERING:** May–Sep | **FRUIT:** pod

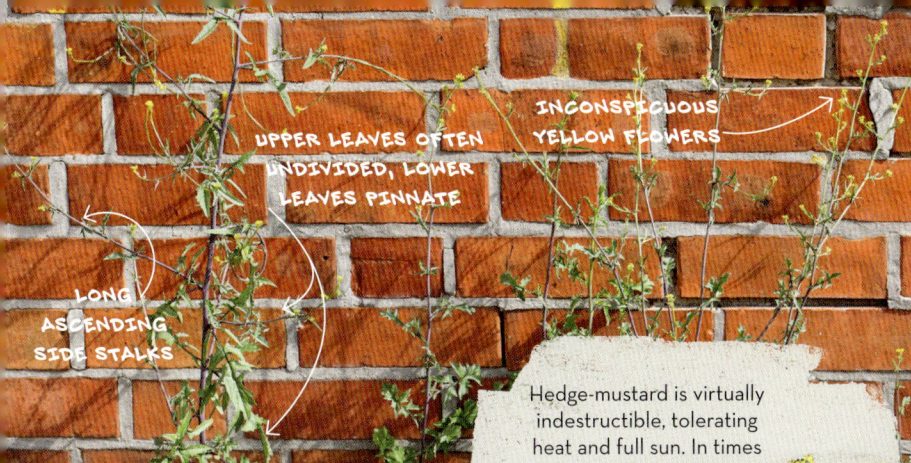

UPPER LEAVES OFTEN UNDIVIDED, LOWER LEAVES PINNATE

INCONSPICUOUS YELLOW FLOWERS

LONG ASCENDING SIDE STALKS

Hedge-mustard is virtually indestructible, tolerating heat and full sun. In times gone by, the dead plants were used as brooms. As the seedpods were still attached, the plant benefited from a very effective dispersal of its seeds.

HEDGE MUSTARD
Cleaning and cooking

Sisymbrium officinale | Brassicaceae

Originating from Europe, western Asia and North Africa, this species is now widespread in cities around the world. It is very tolerant and can grow in many habitats but prefers sunny areas of waste ground. The flowers are an important source of pollen for several species of mining bees. The plant continually grows upwards, constantly forming new flowers.

The leaves are eaten by the caterpillars of at least four different species of butterfly. Also good for human consumption, the young green leaves are excellent in wild salads. Powder ground from the seeds produces an aromatic spice with a mustard-like flavour. Use it sparingly, however, as hedge mustard also contains glycosides that can affect the heart.

DESCRIPTION: erect growth with upwardly protruding arched side stalks, often looks very messy | **GROWTH FORM:** annual | **HEIGHT:** 30–60 cm | **LEAVES:** pinnatifid, lobed, large end section, hairy on both sides | **FLOWERING:** small, inconspicuous, May–Oct | **FRUIT:** thin pods held close to the stem

EXTREMELY LONG SEEDPODS

ARROW-SHAPED UPPER LEAVES

EASTERN ROCKET
Arrowheads for leaves

Sisymbrium orientale | Brassicaceae

As with so many other species, this plant arrived in Britain and Ireland as a contaminant of grain. Favours rough ground, such as vacant lots or dry flowerbeds in cities. The seedpods are very eye-catching, being several centimetres long. Also known by the name Indian Hedge-mustard, this species came to the notice of many in London following the Blitz, when it colonised bomb sites along with various other ruderal species such as Rosebay Willowherb *Chamaenerion angustifolium*.

DESCRIPTION: erect to ascending, often messy looking | **GROWTH FORM:** annual | **HEIGHT:** 10–100 cm | **LEAVES:** pinnatifid below, lobed, large end section, hairy on both sides, as with several of its congeners; however, the upper leaves are strikingly distinct: arrowhead-shaped or even linear | **FLOWERING:** 🍀 Apr–Dec | **FRUIT:** exceedingly long, gently curved pod held out from the stem

FOUR GLOWING YELLOW PETALS

DEEPLY LOBED LEAVES

Before you pick the delicious leaves for a salad or pizza, you should look carefully to see where the plant is growing, and in all instances wash it very throroughly!

PERENNIAL WALL-ROCKET
Rucola at the edge of the road

Diplotaxis tenuifolia | Brassicaceae

This close relative of Garden Rocket *Eruca vesicaria* has a similar characteristic, spicy, nutty rocket flavour – the mustard oils in both species are responsiblc for this. In contrast to salad rocket with its whitish flowers, wild rocket attracts attention with its bright yellow showy blooms. It is very much at home along the edges of large roads, as it can cope with salt and loves light and warmth. Good prerequisites for life in the city!

DESCRIPTION: deep green with whitish waxy coating, branched, fast-growing plant, 2–6 cm long pods | **GROWTH FORM:** herbaceous perennial | **HEIGHT:** 30–80 cm | **LEAVES:** 2–15 cm long, 1–6 cm wide, elliptical to obovate, from corrugated edge to quite deeply lobed with 2–5 cm longish or linear sections | **FLOWERING:** 🌼 May–Oct | **FRUIT:** pod

FOUR CRUMPLED YELLOW PETALS

GREEN-GREY LEAF WITH FROSTED APPEARANCE, UNDERSIDE PALE AND SLIGHTLY HAIRY

YELLOW-ORANGE LATEX SAP

Every seed carries a fleshy appendage that ants like to eat. The ants simply drop the seeds afterwards and so disperse them. Other dispersal agents include snails, shoes, bicycle tyres and cars.

GREATER CELANDINE
Powerful wart remover?

Chelidonium majus | Papaveraceae

Orange-yellow milky latex flows out immediately if the plant is injured. It contains over a dozen alkaloids, including chelidonine, after which Linnaeus named the plant. When he experimented by applying the milky substance on himself, his warts eventually disappeared, but we can only speculate as to whether this was really the latex or just they just got better by themselves. This species is actually a member of the poppy family, and its crumpled petals are a giveaway for this.

DESCRIPTION: orange-yellow sap, four-petalled yellow flowers | **GROWTH FORM:** herbaceous perennial | **HEIGHT:** 30–70 cm | **LEAVES:** alternate, pinnatifid or simple pinnate, wavy, notched margins | **FLOWERING:** Apr–Oct | **FRUIT:** bilobed, longish capsule with ovate seeds with fleshy appendages

BRIGHT YELLOW PETALS WITH CRINKLED EDGES

HAIRY DROOPING BUDS

FEATHERY COMPOUND LEAVES

WELSH POPPY
Sunshine from the hills

Papaver cambricum | Papaveraceae

The long, ribbed seed capsules have flaps that fold back, allowing the hundreds of tiny dark brown kidney-shaped seeds to scatter.

A native plant of shady woodlands and rocky places in the west, this has long been a garden favourite and so is frequently encountered naturalised in towns and cities. The flowers are roughly 5–7 cm across, with four bright yellow (sometimes orange) crinkly petals. This plant is so synonymous with Wales that it forms the logo of the Welsh political party Plaid Cymru. Its pollen-laden, nectar-rich blooms offer an excellent resource for pollinators. The increasing garden escape **California Poppy** *Eschscholzia californica* also has yellow/orange flowers but has much more feathery, grey-green leaves and pointed conical seed capsules.

DESCRIPTION: erect, bright green foliage and eye-catching blooms | **GROWTH FORM:** long-lived perennial | **HEIGHT:** to 60 cm | **LEAVES:** sparsely hairy, stalked, pinnate with pinnately lobed leaflets | **FLOWERING:** ✿ Jun–Aug | **FRUIT:** elongated capsule with 4 to 6 ribs

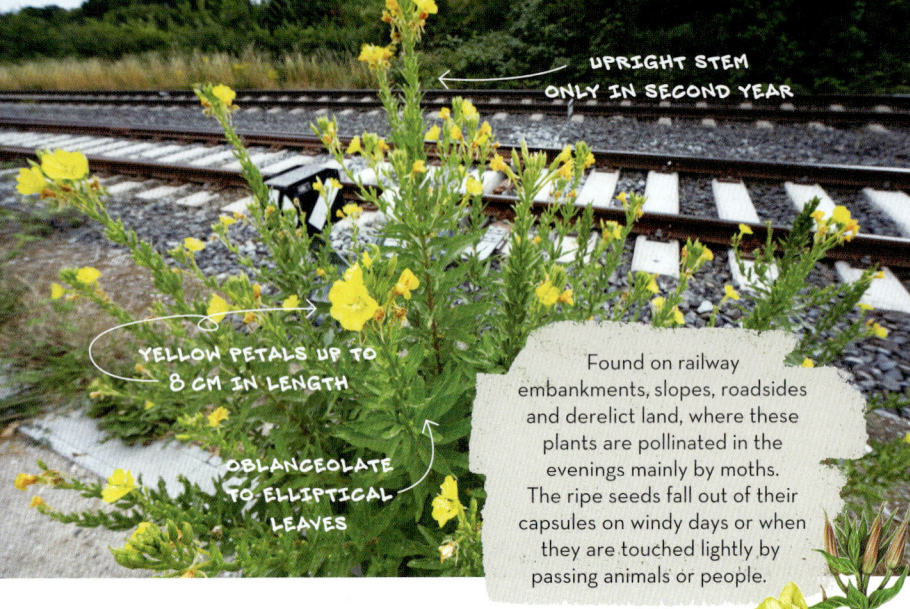

UPRIGHT STEM ONLY IN SECOND YEAR

YELLOW PETALS UP TO 8 CM IN LENGTH

OBLANCEOLATE TO ELLIPTICAL LEAVES

Found on railway embankments, slopes, roadsides and derelict land, where these plants are pollinated in the evenings mainly by moths. The ripe seeds fall out of their capsules on windy days or when they are touched lightly by passing animals or people.

COMMON EVENING-PRIMROSE
Yellow queen of the night

Oenothera biennis | Onagraceae

The fragrant flowers are so named because they unfurl in the evening and bloom until late the next morning. Evening-primrose was brought to Europe from North America as an ornamental plant around 1620. The plant is not only beautiful, but also very tasty. The reddish taproot can be prepared like salsify, and the leaves taste good in a salad or cooked like spinach. The pale yellow flowers are an edible decoration, and a valuable oil is extracted from the seeds. Several species now occur in Britain and Ireland, but this one is the most common.

DESCRIPTION: rosette in first year; upright stem with many big flowers up to 8 cm across in second year | **GROWTH FORM:** biennial | **HEIGHT:** 50–200 cm | **LEAVES:** light green, narrow oblanceolate to elliptical, basal leaves up to 30 cm, stem leaves up to 22 cm in length and up to 5 cm in width | **FLOWERING:** Jun–Sep | **FRUIT:** capsule

STAR-SHAPED YELLOW FLOWERS

FLESHY SEMI-OVATE LEAVES

The seeds are catapulted out of the fruit during rainfall and dispersed by water. The plant is mainly self-pollinating and also propagates vegetatively by means of roots growing out the stems.

BITING STONECROP
Succulent and spicy

Sedum acre | Crassulaceae

Survives long dry periods and still carries on growing. The leaves store water (meaning that this species is known as a succulent) and it takes a long time for the water supply to be displaced from the leaf cells. This allows stonecrop to grow, flower and spread in dry paved areas. Biting

Stonecrop is a popular plant for green roofs and its showy reddish stems of pretty yellow flowers are popular with insects. The species epithet *acre* denotes that this species tastes bitter, as does 'biting'; however, it should not be consumed as it can cause sickness.

DESCRIPTION: fleshy leaves and yellow flowers | **GROWTH FORM:** herbaceous perennial | **HEIGHT:** 3–15 cm | **LEAVES:** alternate semi-ovate scale-like leaves | **FLOWERING:** 🌼 cluster with 1 to 7 flowers, Jun–Aug | **FRUIT:** follicle

DOUBLE-EDGED STEMS

YELLOW FLOWERS WITH MANY STAMENS

LEAVES WITH TRANSLUCENT GLANDS

PERFORATE ST JOHN'S-WORT

Lifts the spirits in more ways than one

Hypericum perforatum | Hypericaceae

Named after the typical date of flowering: 24 June is St John's Day, the birthday of John the Baptist – close to the summer solstice (between 20 and 22 June). It is traditionally around this time that St John's-wort begins to bloom, with the flowering period now being pushed further and further forward, but the name remains. Perforate St John's-wort has a double-edged, pith-filled stem, allowing us to distinguish it easily from the other St John's-wort species (which are much less likely to occur in urban areas anyway).

DESCRIPTION: Leaves appear to be perforated, crushed flowers produce red colouring on hands, double-edged, pithy stems | **GROWTH FORM:** herbaceous perennial | **HEIGHT:** 30–60 cm | **LEAVES:** alternate, simple leaves with translucent glands | **FLOWERING:** 🌼 in messy clusters, Jun–Aug | **FRUIT:** dehiscent capsule

RED FINGERS FROM YELLOW FLOWERS

If you crush the flowers between your fingers, you will get red fingertips. This is due to the presence of the compound hypericin, after which the genus was named. This reddish liquid is associated with the blood of the beheaded St John the Baptist and gives the plant its name.

Perforatum, that is 'punctured', is the scientific name of this species of St John's-wort – here you can see why: the leaves appear to have tiny holes.

HERB AGAINST DEPRESSION

Its ability to counter low moods is fairly well known, and St John's-wort can be bought in the form of a tea or capsules from many pharmacies. Perhaps, then, it is no bad thing that it is spreading: there are more and more studies showing a high incidence of mental illness in cities, especially in urban areas without greenery. So please let the plant grow, and don't hesitate to campaign for more native greenery in our urban areas – it's good for us all. However, we would like to point out that tea made from the herb does not help against severe depression, as many studies have shown.

LOTS OF POLLEN, NO NECTAR

Many different bee species are attracted to the pretty, orange-coloured pollen grains, even though St John's-wort does not provide nectar. To recharge their batteries for collecting this pollen, bees and flies have to visit other flowering species in between.

Dispersal strategies: the small seeds are moved by the wind, but also carried by insects. Perforate St John's-wort can propagate vegetatively too, by way of creeping rhizomes.

If you crush the buds and flowers between your fingers, a red dye is excreted.

HEDGEHOG-LIKE FRUITS

SMALL BRIGHT YELLOW FLOWERS ON UPRIGHT STEMS.

EVERGREEN LEAF ROSETTE

WOOD AVENS
Clingy seedheads resembling hedgehogs

Geum urbanum | Rosaceae

Spicy roots that smell of cloves are a distinctive feature of this widespread species that you can find everywhere in the urban environment. The clue is in its scientific name *urbanum*, which – despite the English monicker of 'wood' – indicates that it likes to grow in and around human settlements. Its turnip-shaped rhizomes contain gein (pronounced 'ge-in'), a glycoside which, during drying, is extracted for the intensely clove-scented

DESCRIPTION: evergreen rosette made up of odd pinnate, hairy leaves with petioles | **GROWTH FORM:** herbaceous perennial | **HEIGHT:** 30–120 cm | **LEAVES:** stem leaves trifoliate with short petioles, alternate | **FLOWERING:** 🌼 May–Sep | **FRUIT:** achenes with burrs

essential oil eugenol. The young leaves, shoots and flowers are edible, and some people even use the roots, fresh or dried, as a spice. The rhizomes were also used medicinally in the past, for example against skin diseases, fever and inflammation, for stomach and digestive problems, and even as an aphrodisiac. The roots can also be added to toothpastes and mouthwashes as they prevent inflammation in the mouth. An extract from the roots was used to refine herbal liqueurs, wine or beer. It was held that the powdered roots could drive away witches and devils.

LIGHT AND COOLNESS

These are the conditions that Wood Avens likes most of all, thus preferring semi-shaded woodland edges and nutrient-rich soil. But it also does well in towns, often on urban fallow and shady waste ground, as well as being a prolific weed in gardens.

SMALL FLOWERS WITH TINY VISITORS

Flies, hoverflies and beetles are the most frequent visitors to the small yellow flowers, which sit on sparse stems. Other insects rarely come to the flowers. The plant is also a foodplant for the caterpillars of several species of moth.

BALLS WITH HOOKS

The ripe fruits form small spherical heads with hooks pointing outwards

The fruits of Wood Avens almost look like small hedghogs. Thanks to their hooks, the seeds stick to fur or clothing and thus get dispersed very effectively.

that look like little hedgehogs. The hooks attach themselves to passing animals or even to human clothing, thus dispersing the seeds.

After flowering it is possible to identify Wood Avens most readily. The spherical upright fruiting heads make them unmistakable.

DELICATE, MOSTLY REDDISH-TINGED RUNNERS

FIVE BROAD, BRIGHT YELLOW PETALS

HAND-SHAPED LEAVES

In urban areas we tend to find these delicate creepers around trees, on roadsides and the edges of pavements rather than in cracks in paving stones.

CREEPING CINQUEFOIL
Fast-growing ground cover

Potentilla reptans | Rosaceae

Resembling tiny hands, its five-lobed leaves have given the species another name – five-finger grass. Gardeners do not welcome it, as its runners, sometimes up to a metre long, spread out from the taproot, which is itself up to 45 cm long, covering flowerbeds and vegetable plots with a dense carpet of leaves and yellow flowers. The young parts are loved by rabbits and guinea pigs. Over 20 species of bee collect pollen from its flowers and the seeds are dispersed by ants.

DESCRIPTION: creeping plant with hand-shaped leaves on often reddish tinged runners of up to a metre in length, each with multiple rooting leaf nodes | **GROWTH FORM:** herbaceous perennial | **HEIGHT:** 10–20 cm | **LEAVES:** alternate, with five lobes, finely serrate | **FLOWERING:** 🌼 Jun–Aug | **FRUIT:** nutlets or achenes

GLOSSY GOLDEN-YELLOW FLOWERS

GROOVED PEDICEL

THREE LEAFLETS

This species is found in clumps, rarely as an individual plant because it produces large numbers of young plants through its very long overground runners and its prodigious seed output.

CREEPING BUTTERCUP

Leafy runners stretching for metres

Ranunculus repens | Ranunculaceae

This species contains less of the toxins known as protoanemonins than other buttercups and is therefore categorised as only slightly toxic. The ability to penetrate compacted soils is one of its advantages and so it can be found wherever the ground is dense and hard, such as trampled areas, while it also favours damp places in particular. It can be quite thuggish in gardens and allotments, rapidly forming a dense carpet if allowed. The flowers are a good source of nectar and pollen, especially for small beetles, moths and ants.

DESCRIPTION: petals carry nectar at their base in a nectary | **GROWTH FORM:** herbaceous perennial | **HEIGHT:** 15–40 cm | **LEAVES:** alternate with three leaflets, with a petiole on the middle leaflet | **FLOWERING:** 🌼 with a grooved pedicel, May–Aug | **FRUIT:** achene

UPRIGHT SEED CAPSULES

TYPICAL TRIFOLIATE LEAVES

The seeds are expelled from the erect fruits for distances of over a metre, thereby reaching every corner of our towns – just see for yourself by touching them!

TINGED RED

PROCUMBENT YELLOW-SORREL
Sun-worshipper

Oxalis corniculata | Oxalidaceae

In the nineteenth century this pretty little plant was probably brought to Northern Europe from the Mediterranean. It likes warmth but does not tolerate frost and overwinters as a seed. It copes very well with drought conditions, however, and protects its leaves from intense light by its dark red colour. The flowers only open in direct sunlight. If a yellow-sorrel (there are several species present in Britain and Ireland now) has worked its way into the flowerpots on your balcony (ants carry the seeds), be pleased: the leaves taste refreshingly tart – like sharp apple peel – and are tasty in salads.

DESCRIPTION: creeping runners form loose reddish-brown cushions | **GROWTH FORM:** annual or perennial | **HEIGHT:** 10–50 cm | **LEAVES:** alternate petiolate leaves, each divided into three heart-shaped leaflets | **FLOWERING:** 🌼 Jun–Sep | **FRUIT:** capsule

PUNGENT SMELL

SHOOTS BRANCH MANY TIMES

GLANDULAR-HAIRY

TOMATO
Escape from a fast-food restaurant?

Solanum lycopersicum | Solanaceae

Tomatoes can reproduce through runners. In urban areas they originate from seeds that have been thrown away.

The tomato germinates easily, but rarely flowers in locations with a lot of footfall. Tomato flowers are pollinated by the wind or by bumblebees with their fast wingbeats in so-called buzz-pollination. Tomatoes originally came from the Americas, probably arriving in Europe with Columbus. Nowadays this species is cropping up more and more in the urban environment, having been famed among botanists for some time as a frequent denizen on sewage works (its seeds are happy passing through the human gut).

DESCRIPTION: glandular hairs and a strong smell of tomato over the whole plant | **GROWTH FORM:** annual or biennial | **HEIGHT:** 20–200 cm | **LEAVES:** alternate, interrupted odd pinnate | **FLOWERING:** 🌼 fused yellow petals, stamens fused to the anthers, May–Aug | **FRUIT:** berry (yes, tomatoes are, botanically speaking, berries)

LEAFLESS HOLLOW STEM

YELLOW RAY FLORETS

WHITE LATEX

LEAVES IN BASAL ROSETTE

Dandelion is the best-known plant to use wind for its seed dispersal. The calyx of each single flower in the flowerhead becomes a small parachute. Who doesn't enjoy blowing the seeds of a Dandelion clock into the wind?

DANDELION
Time for a wee?

Taraxacum officinale | Asteraceae

Dandelions love nutrient-rich, deep soils, but we can also find small rosettes of leaves in dry pavement cracks. They produce a lot of pollen and are popular with bees of all kinds. However, because the pollen does not contain all the essential amino acids, insects cannot feed on this alone. An old name for Dandelion is 'piss-a-bed', due to its diuretic qualities. There are in fact hundreds of different micro-species of Dandelion, most obviously distinct in the leaf shapes – see how many versions you can spot near your home?! There are worse things than flinging a few leaves into a salad, but perhaps better from a meadow rather than the pavement.

DESCRIPTION: Leafless, hollow stem with milky latex sap | **GROWTH FORM:** annual | **HEIGHT:** 10–30 cm | **LEAVES:** pinnatifid leaves in a basal rosette | **FLOWERING:** 🌼 yellow ray florets, Apr–Jul | **FRUIT:** achenes with parachutes for seed dispersal

YELLOW RAY FLORETS

MULTIPLE BRANCHING STEMS

LATEX

The pappi (seeds with 'parachutes') are carried a long way by the wind, but also attach themselves to animals. All animals that eat Dandelions also like Smooth Hawks-beard.

SMOOTH HAWK'S-BEARD
Parachuting seeds

Crepis capillaris | Asteraceae

This plant not only grows frequently in nutrient-poor pastures and meadows, but also in weedy fields, fallow land and waste tips. It often adorns the edges of paths and walls with its yellow flowers. Mining bees love the small yellow flowerheads, which consist only of ray florets. Self-pollination appears to be common as well. Smooth Hawk's-beard leaves can be used in salads just like Dandelion. If your plant is hairy and big, consider Beaked **Hawk's-beard *Crepis vesicaria*** or **Rough Hawk's-beard *C. biennis*** too.

DESCRIPTION: leaves similar to dandelion, multiple flowerheads; stems contain latex sap | **GROWTH FORM:** perennial | **HEIGHT:** 10–70 cm | **LEAVES:** most leaves are pinnatifid and basal, those on the stem smaller with pointed lobes | **FLOWERING:** only ray florets, May–Oct | **FRUIT:** achenes with parachutes, like the Dandelion

CREAMY YELLOW-ORANGE LATEX

NORTH–SOUTH ORIENTATION OF LEAVES

VERTICALLY POSITIONED LEAF EDGE

PRICKLY MAIN VEIN ON UNDERSIDE

PRICKLY LETTUCE
Behaves like a compass

Lactuca serriola | Asteraceae

The stiff leaves of Prickly Lettuce generally point north–south; this peculiarity is the reason why the species is also known as 'compass lettuce'. The compass position is an adaptation to heat and very strong sunlight: the leaf blade is vertical and the narrow side points in a north–south direction, protecting it from intense solar radiation. However, the orientation of the leaves changes depending on the incidence and intensity of the rays. The movement with which plants react to the stimulus of sunlight is called phototropism. You can also observe this in sunflowers, for example.

DESCRIPTION: main vein on the underside of the leaf is prickly; leaves are held in the vertical plane; stem emits latex when touched or broken | **GROWTH FORM:** annual or biennial | **HEIGHT:** 60–230 cm | **LEAVES** alternate with pinnatifid or lobed prickly leaves | **FLOWERING:** ☀ 15–25 yellow ray florets, Jun–Aug | **FRUIT:** achenes with stemmed feathery parachutes for wind dispersal

THE LEATHERY LEAVES ARE ROUGH AND PRICKLY, and the plant favours growing by walls and fences. Prickly Lettuce also spreads along roadsides or railway tracks. Take a look out of the train window if it happens to stop, and there's a good chance that you'll see this tall plant with its dark green colouring and small yellow flowers in summer.

Likes to grow along fences and the edges of paths.

The hairy tufts (pappus) on the seeds of Prickly Lettuce have stems. The seeds are very light and thanks to this 'flight aid' are highly efficient in their wind dispersal.

CATERPILLARS OF THE SMALL RANUNCULUS and other species of moth like to feed on this succulent plant with its thick milky sap (this latex gives its name to the genus *Lactuca*). A tip if you ever come across a caterpillar on a plant: with the iNaturalist app, for instance, you can identify animals and plants using photos and experts will check the identification. This works well for caterpillars.

BITTER PLANT

Especially after flowering, the edible Prickly Lettuce contains many bitter substances. It is therefore not very popular in salads. However, the plant has found its niche in smoothie recipes together with other fruit and vegetables, as the bitter substances are healthy. Moreover, the leaves contain vitamins and iron too.

BLUE-GREEN LEAVES, OFTEN LOOKS MATTE

FLOWERHEADS CLOSE AT MIDDAY

REDDISH-TINGED STEMS AND LEAF VEINS

SMOOTH SOW-THISTLE
Medieval vegetable and medicinal herb

Sonchus oleraceus | Asteraceae

As long ago as the middle ages, the sow-thistle came to us from the Mediterranean as a pot herb. Today it can be found in temperate regions all over the world. We still love this traditional vegetable when we are wild camping: it is just a matter of collecting and steaming the leaves. They contain plenty of vitamin C, as well as iron and other minerals. The rhizomes can be dried and used to make flour to thicken sauces. Like dandelions, the

DESCRIPTION: serrate, prickly, but matte-looking leaves with a reddish edge | **GROWTH FORM:** annual | **HEIGHT:** 30–100 cm | **LEAVES:** alternate, clasping the stem, serrate but thornless and hairless; pointed lobes at the base of the leaves | **FLOWERING:** ✹ many yellow ray florets, no disc florets, Jun–Oct | **FRUIT:** achenes with pappus, significantly denser than in the dandelion

leaves, hollow stems and flowers are eaten by hares and rabbits, hence one of its other names, hare's thistle.

OLD MEDICINAL HERB

Sow-thistle was already in use as a medicinal plant in the Middle Ages because it contains the compound taraxasterol. This active ingredient exists in high concentration in the milky sap and has an anti-inflammatory effect; it can be used to treat skin diseases such as acne. Whereas in the medieval period sow-thistle was used in many different ways for internal and external applications, today we rarely harness its various natural benefits.

As the Common Sow-thistle has pappus hairs, the seeds are dispersed efficiently by the wind. Dispersal by insects and birds has also been observed – this is facilitated by the nourishing seeds which contain oil.

This species has a knack for squeezing into even the smallest of cracks and can turn up in some odd places.

CLOSED AT MIDDAY

As with many composite plants that only have ray florets, the flowerheads close at noon. Fewer wild bees and flies visit its flowers than they do those of the Dandelion. Nevertheless, it seems to benefit from pollination, because if pollinators are excluded (by putting little bags over the flowers to make them inaccessible), fewer seeds develop.

SMALL YELLOW FLOWERS

VERY DELICATELY BRANCHED INFLORESCENCES

LATEX

LEAVES WITH LARGE TERMINAL LOBE

WALL LETTUCE

Enhances old walls, even in the shade

Mycelis muralis | Asteraceae

Beautiful pappi form from the five 'parachutes' with their dark brown seeds. They are not only carried along by the wind, but also by rainwater flowing down walls.

Wind and water help Wall Lettuce to spread successfully. This plant grows particularly well on partially or fully shaded walls. Its yellow flowers have few ray florets. As with other members of the lettuce family, it has a milky sap. Hoverflies, sweat bees (*Lasioglossum*) and mining bees (*Panurgus*) are regular visitors. However, self-pollination also works well and the plant is not necessarily dependent on these flower-visiting insects. This species is very pleasing in form and colour.

DESCRIPTION: loose panicle with only five ray florets in each flowerhead | **GROWTH FORM:** herbaceous perennial | **HEIGHT:** 60–80 cm | **LEAVES:** alternate, hairless, lyre-shaped to pinnatifid, undersides bluish-green | **FLOWERING:** 🌼 Jul–Aug | **FRUIT:** achenes with pappus

DEEPLY DIVIDED LEAVES

RICH YELLOW FLOWERHEADS

COMMON RAGWORT
Not poisonous for the Cinnabar moth

Jacobaea vulgaris | Asteraceae

Due to the alkaloid content, all parts of Common Ragwort are poisonous – for humans and many animals, famously livestock. But the orange-and-black-ringed caterpillars of the beautiful day-flying Cinnabar moth feed on the plant, absorb the alkaloids and themselves become inedible to predators. The alkaloids are also found in the honey when honeybees collect nectar from Ragwort, which is why urban bee-keepers are less than enthusiastic about this species and its congeners.

DESCRIPTION: numerous yellow flowerheads with ray and disc florets | **GROWTH FORM:** biennial | **HEIGHT:** 30–100 cm | **LEAVES:** alternate, pinnatifid, bare leaves, sometimes with hairs resembling cobwebs | **FLOWERING:** 🌼 yellow disc florets and long ray florets, Jul–Sep | **FRUIT:** achenes with pappus

LOOSE, MANY-BRANCHED INFLORESCENCE

SHOWY, BRIGHT YELLOW FLOWERS

LONG, NARROW, DARK GREEN LEAVES

As attractive as this plant may be, it is not always very welcome. It is poisonous, often forms large populations and in some places displaces native species.

NARROW-LEAVED RAGWORT

New migrant rapidly gaining ground

Senecio inaequidens | Asteraceae

The genus name *Senecio* means 'old man' and this genus was probably so called because of the whitish crowns of pappus hair on the seeds, which are reminiscent of an old man's hair. This species has been found in European port cities since the nineteenth century, having been imported by accident with wool from South Africa. However, it is only since the 1970s that it has spread along motorways and railway lines throughout the continent and is becoming ever-more common. The light seeds are carried along by the wind.

DESCRIPTION: strongly branched plant that likes to form large, conspicuous stands in warm, dry locations | **GROWTH FORM:** perennial | **HEIGHT:** 20–60 cm | **LEAVES:** dark green, linear to narrow-lanceolate, clasping the stem at the bottom, only 1–5 mm wide, approx. 6 cm long, finely toothed | **FLOWERING:** ✿ Aug–Oct | **FRUIT:** achenes with pappus like the Dandelion

PLANTS PRETTY MUCH HAIRLESS

FLOWERS USUALLY SLIGHTLY LARGER, BRIGHTER AND NEATER LOOKING THAN COMMON RAGWORT

GENERALLY A STOCKIER PLANT THAN COMMON RAGWORT

OXFORD RAGWORT
Gown goes to town

Jacobaea squalidus | Asteraceae

This species has a most interesting history: first recorded in the wild in the late eighteenth century, it escaped from Oxford Botanic Garden where it had derived from a hybrid of species occurring on the slopes of Mount Etna, Sicily. It is actually now declining, perhaps due to more assiduous weeding of streets, but is still a pretty common sight in cities, along railway lines and on dry waste ground. The flowers often appear larger and more golden yellow than those of Common Ragwort, but plants are very variable (betraying their hybrid origins). The most reliable character for identification is the conspicuously black-tipped phyllaries on the sides of the flowerheads (see the cover of this book where they can clearly be observed on the buds).

DESCRIPTION: numerous yellow flowerheads with ray and disc florets | **GROWTH FORM:** Annual or perennial | **HEIGHT:** to 75 cm | **LEAVES:** alternate, pinnatifid, with a pointed terminal lobe, usually hairless | **FLOWERING:** 🌼 yellow disc florets and long ray florets, Apr–Dec | **FRUIT:** achenes with pappus

BLACK TIPS ON PHYLLARIES

NODDING FLOWERHEADS, ONLY DISC FLORETS

PINNATIFID LEAVES

The seeds are endowed with hairy tufts like a dandelion clock and are dispersed by the wind. The fruits exude a sort of slime which can stick to shoes and tyres.

GROUNDSEL
Good flyer with a colourful past

Senecio vulgaris | Asteraceae

Spreads rapidly in urban areas and occasionally even flowers in winter. Groundsel is mainly frequented by hoverflies but can also self-pollinate if necessary. The plant contains pyrrolizidine alkaloids and is therefore poisonous. This species is a common weed in fields in Europe and the USA.

It has been controlled with herbicides for many years and has already developed resistance to some chemical agents. In the early modern period, the seeds were gathered and sold as canary food by 'groundsel-men'.

DESCRIPTION: flowerheads with nodding disc florets, black-tipped bracts at the base | **GROWTH FORM:** annual | **HEIGHT:** 10–30 cm | **LEAVES:** alternate, pinnatifid to irregularly pinnate | **FLOWERING:** ✽ yellow-green disc florets, usually no ray florets, Feb–Nov | **FRUIT:** achenes with pappus like the dandelion

FLOWERHEADS WITHOUT RAY FLORETS

FEATHERY, HAIRLESS LEAVES.

PLEASANT PINEAPPLE/CHAMOMILE SCENT

When it rains the seeds swell and become slimy. They stick to shoes and bicycle tyres. When a car tyre picks up a seed, it can be transported over very long distances.

PINEAPPLEWEED
Sticky seeds to aid dispersal

Matricaria discoidea | Asteraceae

Introduced from North America and North Asia, we now frequently find this chamomile-like species in towns and villages. Unlike chamomile, it does not have ray florets and was considered a rather a curiosity for this before its massive global expansion. Pineappleweed is fragrant – smelling pleasantly of pineapple, as the name suggests – but not as strongly scented as true chamomile. Its essential oil has a different composition to that of true chamomile and is not as important in naturopathy.

DESCRIPTION: flowerhead without ray florets, aromatic scent | **GROWTH FORM:** annual | **HEIGHT:** 5–30 cm | **LEAVES:** alternate, pinnately dissected, hairless | **FLOWERING:** 🌼 yellow-green disc florets, no ray florets, Jun–Aug | **FRUIT:** achene without pappus

NO LEAVES WHILST FLOWERING

THE FLOWERHEAD DROOPS DOWN AFTER POLLINATION

DOWNY STEMS WITH SCALES

COLTSFOOT
Gets rid of coughs in any weather

Tussilago farfara | Asteraceae

The bright eye-catching yellow of Coltsfoot flowers is particularly uplifting whey appear in late winter, sometimes even poking through snow. Among its many local names are Cleat, Clote, Clatterclogs, Dummy-weed, Yellow Trumpets, Son-afore-the-Father and Poor Man's Baccy – indeed, the leaves used to be made into a herbal tobacco. A pioneer species of disturbed or rough ground, Coltsfoot was always going to be well placed to venture into town, where it can appear on walls, in corners or in vacant lots.

DESCRIPTION: no green leaves during flowering; flowerheads singly on thick downy stems covered in scale-like bracts | **GROWTH FORM:** herbaceous perennial | **HEIGHT:** 7–30 cm | **LEAVES:** basal, heart-shaped to rounded with a downy underside and blackish tips | **FLOWERING:** 🌼 very narrow bright yellow ray florets, few disc florets, Mar–Apr | **FRUIT:** achene with pappus

COLTSFOOT OR BUTTERBUR?

Many people ask themselves this question when only the leaves of the plant are visible. However, as Butterbur *Petasites hybridus* prefers moist and shady habitats, it is rarely found in the city. If you look closely, you can in fact easily distinguish the two species when vegetative: Coltsfoot has blackish tips on its leaves, butterbur does not. When you see the two plants in flower, however, there is no confusion possible.

SPECIAL FLOWERHEADS

There are both male and female flowers within a single flowerhead: the numerous very narrow ray florets are female, that is, they only have carpels and no stamens. The comparatively few disc florets, on the other hand, are all male and only have stamens.

RUNNERS OF UP TO SEVERAL METRES

Coltsfoot not only uses the wind to spread its seeds but can also colonise new habitats vegetatively. It forms long root runners from its rhizome that burrow their way around stones and tree roots until they find a favourable place to grow.

CURES COUGHS

Part of the scientific name *Tussilago* reveals that this plant can help against coughs (think 'antitussive'). However, as Coltsfoot also contains poisonous alkaloids, it is not advisable to use the wild plant as a medicinal herb. Special alkaloid-free cultivars have been developed that can be employed for medicinal purposes.

When the pollinated flower has withered, the seedhead hangs down, nodding in the wind. This helps the dispersal of the seeds.

Coltsfoot (above) and Butterbur (below)

PYRAMID-SHAPED INFLORESCENCES WITH MANY PANICLES

YELLOW FLOWERHEADS WITHOUT RAY FLORETS

RHIZOMES PROPAGATING NEW PLANTS

CANADIAN GOLDENROD
Attractive garden escape

Solidago canadensis | Asteraceae

It was only in the nineteenth century that this species started to spread rapidly over Europe and become invasive, although it had already arrived from North America in 1648 at the botanical gardens of Paris. It was cultivated as an ornamental plant and for bee pasture and is still obtainable in garden centres. Nature conservationists are less enthusiastic about it and are now introducing measures to control the plant, as various studies have shown that it can displace native flora.

VERY POPULAR with wild bees and honeybees and, together with Ivy *Hedera helix*, when present it is

---

DESCRIPTION: pyramid-shaped panicles with numerous small yellow flowerheads | **GROWTH FORM:** herbaceous perennial | **HEIGHT:** 50–250 cm | **LEAVES:** alternate, lanceolate, toothed at front | **FLOWERING:** ✿ minute yellow ray and disc florets, Aug–Sep | **FRUIT:** achenes

The many airborne seeds and its rhizomes allow Canadian Goldenrod to spread rapidly. Unlikely to be found in pavement cracks, it tends to occur on waste and building material tips or along railway lines.

the plant that is most frequented by insects in urban areas in late summer and early autumn. Interestingly, it has actually never been clearly established whether it is indeed good for bees to have a large supply of flowers in autumn. In late summer and autumn, there are fewer flowers in natural ecosystems, and native bee species have adapted to this. In cities, on the other hand, non-native plants bloom well into late autumn and even winter. Thanks to the good food supply, bee populations here remain strong until later in the year when they may suddenly be surprised by frost and not have sufficient supplies to last the winter.

DYE PRODUCTION

In the past a yellowish dye was extracted from Canadian Goldenrod. The yellow colour is mainly due to the flavonoids quercetin and astragalin.

Canadian Goldenrod often colonises waste tips or derelict land.

TERMINAL PANICLES

SMALL PALE FLOWERHEADS

DOWNY WHITE UNDERSIDES TO LEAVES

Wind, rain, people and – despite their small size – also birds help spread thousands of tiny seeds.

MUGWORT
The hay fever plant

Artemisia vulgaris | Asteraceae

Although it may be a weed of root crops, Mugwort can actually be easily removed from the soil by digging up the root. The plant is highly floriferous and may trigger hay fever when its pollen is spread by the wind. However, Mugwort was held to have magical properties, able to ward off evil. If you look carefully at the leaves, in some areas you may be lucky enough to find the caterpillar of a rare moth, the Wormwood *Cucullia absinthia*, with its green-and-white-striped pattern.

DESCRIPTION: leaves green on top and downy white underneath, weak aromatic smell when rubbed | **GROWTH FORM:** herbaceous perennial | **HEIGHT:** 60–250 cm | **LEAVES:** alternate and deeply divided | **FLOWERING:** 🌼 Jul–Nov | **FRUIT:** single-seeded achene

TIGHTLY BUNCHED GLOBULAR YELLOW FLOWERHEADS

The seeds are dispersed by rain, and in urban areas stick to the soles of shoes. Roots may also develop on shoots, allowing vegetative propagation.

AS MEDICKS ARE IN FACT CLOVERS, THEIR LEAVES HAVE THREE LEAFLETS

BLACK MEDICK
Explodes when disturbed

Medicago lupulina | Fabaceae

Everywhere, but mostly overlooked is an excellent description for this small yellow clover whose flowers are somewhat reminiscent of hop blossom. It is a pioneer along roadsides and in dry meadows with calcareous and clay soils. It can be distinguished from other clover species with small yellow flowers by its leaves: if the midrib at the top of the leaf sticks out as a small point, you are looking at Black Medick. This plant also has a cunning trick, namely, if an insect lands on the flowerhead, the stamens shoot out and explosively discharge their pollen onto the unsuspecting visitor.

DESCRIPTION: leaflets with short sharp tip; black wrinkly fruits | **GROWTH FORM:** annual or biennial | **HEIGHT:** 15–60 cm | **LEAVES:** alternate and compound, each with three leaflets – middle one with longer petiole | **FLOWERING:** May-Oct | **FRUIT:** single-seeded pod

VIOLET, WHITE AND YELLOW PETALS

DARK VEINS

DEEPLY PALMATELY LOBED STIPULES

Whilst today only rarely found in arable fields, Wild Pansy is common in urban habitats on grassy and fallow areas and embankments. It favours sandy nutrient-poor soils in which its roots extend down to depths of up to 45 cm.

WILD PANSY
Dazzling splashes of colour

Viola tricolor | Violaceae

Also known by the traditional name of Heartsease, this native species was formerly a common arable weed (and indeed it still occurs occasionally in crop fields). Many examples in urban areas may actually be cultivated varieties which have hopped out of planters or hanging baskets – careful looking can sometimes find the source nearby/above. Wherever they appear, these charming flowers are always a delight, even with detritus and dog hairs tangled in the leaves. Wild Pansy can also serve as food for the caterpillars of several different butterfly species.

DESCRIPTION: delicate flowers with violet, white and yellow petals and dark veins | **GROWTH FORM:** annual but also very rarely herbaceous perennial | **HEIGHT:** 5–15 cm | **LEAVES:** lower leaves heart-shaped to ovate, notched, on an upright usually branched stem | **FLOWERING:** 🌼 Apr–Sep | **FRUIT:** capsule

LOOSE RACEMES OF BRIGHT YELLOW FLOWERS

BUSHY GROWTH HABIT

PEA GREEN LEAVES WITH ROUNDED LOBES

YELLOW CORYDALIS
A fanfare from the masonry

Pseudofumaria lutea | Papaveraceae

Native to the foothills of the southern and central Alps, this species has been grown in gardens in Britain since the late 1500s (it is still uncommon in Ireland) and is a very frequent escape, swiftly colonising new areas – especially old walls in towns and cities. Formerly also known as Yellow Fumitory, which reveals it as a close cousin of Common Fumitory (see page 95). Its attractive leaves and uplifting sprays of bright yellow flowers ensure that it is always a welcome sight. (Look out for **White Corydalis** *Pseudofumaria alba* too, which has white flowers tipped with dark yellow.)

DESCRIPTION: stems spreading, erect or dangling; feathery leaves and distinctive clusters of bright yellow flowers | **GROWTH FORM:** perennial | **HEIGHT:** 10–30 cm | **LEAVES:** bipinnate or tripinnate | **FLOWERING:** Apr–Oct | **FRUIT:** rounded capsule containing several seeds

ORANGE-YELLOW
BOSS

LONG FLOWER
SPUR

SESSILE, LINEAR
OR LANCEOLATE
LEAVES

The roots of Common Toadflax extend up to a metre under the ground. It grows in warm places such as stony embankments, waste tips and dry paths and has adapted well to the urban environment.

COMMON TOADFLAX
Its pollinators need to be strong

Linaria vulgaris | Plantaginaceae

Despite its name, Toadflax is actually a cousin of the snapdragons in our gardens; it is not related to the blue-flowered flax plant, even though the leaves are similar. The upper lip is divided into two lobes that are folded upwards, while the lower lip, also consisting of two lobes, folds downwards. In between there is an inflated orange-yellow patch that seals the entrance to the throat of the corolla. As a result, insects have to force open the flower to reach the nectar. However, bumblebees save themselves this effort and prefer to puncture the spur from the outside – stealing nectar and giving the plant nothing in return.

DESCRIPTION: pale yellow flower with long spur whose throat is sealed with an orange-yellow swelling | **GROWTH FORM:** herbaceous perennial | **HEIGHT:** 20–75 cm | **LEAVES:** sessile, 2–5 cm in length, 1–1.5 mm in width, linear to lanceolate | **FLOWERING:** clustered spires of flowerheads, Jun–Oct | **FRUIT:** capsule with seeds 2–3 mm long

TWO SETS, EACH OF FOUR
DELICATE RED PETALS, MANY
STAMENS

THIN STEM WITH
BRISTLY HAIRS

ALTERNATE PINNATIFID
LEAVES

COMMON POPPY
Flashes of scarlet

Papaver rhoeas | Papaveraceae

As a typical field and wayside weed, the Common Poppy has accompanied human settlements since the Neolithic. It copes well with dry, warm stony soils, and always provides a welcome splash of colour in our urban environment. Its flowers do not produce nectar, but they do provide huge amounts of pollen. Each of its pretty seed capsules scatters up to 5,000 tiny seeds. There is even a patent for a salt cellar modelled on

Although the Poppy looks delicate, it is a robust, deep-rooted plant with roots up to one metre long, which grows in cities mainly on rubble, along paths or railway tracks. It is, however, sensitive to trampling.

the poppy seed capsule – probably the world's first example of bionics. **Long-headed Poppy** *Papaver dubium* can also be frequent in urban areas; as the name suggests, its seedheads are elongated, and the leaves are more feathery.

DESCRIPTION: flowers appear delicate and crumpled on long thin, hairy stems, with white milky sap | **GROWTH FORM:** annual | **HEIGHT:** 30–90 cm | **LEAVES:** bristly hairs, approx. 15 cm in length, lanceolate, simple pinnate or bipinnate with indented to sharply serrated edge | **FLOWERING:** ❀ May–Jul | **FRUIT:** capsule

FLOWERS ON LONG PEDICELS

DENSELY HAIRY

PALE PINK OR DARK PURPLE FLOWERS WITH FOUR PETALS

This species is easily confused with Broad-leaved Willowherb *Epilobium montanum*, which also occurs in urban areas. The latter has hairless leaves with petioles.

HOARY WILLOWHERB
A modest little bloom

Epilobium parviflorum | Onagraceae

From the canary islands to China, this graceful, softly downy plant can be found throughout the temperate zone. It favours moist, nutrient-rich locations and therefore does not grow everywhere in cities. However, you will come across it from time to time in semi-shaded corners. The reddish-tinged parts that are above ground die off in autumn, while the rhizome sprouts again the following spring. An extract of the plant has an antibacterial and anti-inflammatory effect. Leaves and flowers are edible. Several other willowherb species are also common in towns and cities.

DESCRIPTION: fleecy-haired seeds are dispersed from slim longitudinally dehiscent capsules | **GROWTH FORM:** herbaceous perennial | **HEIGHT:** 30–80 cm | **LEAVES:** alternate, sessile or rarely with a very short stalk, slightly serrate, hairy; narrow elliptical to lanceolate; max. 12 cm long, 3 cm wide | **FLOWERING:** ✽ Jun–Sep | **FRUIT:** capsule

LEAVES IN WHORLS AROUND THE STEM

FOUR PALE PINK PETALS WITH SHARP POINTS

FIELD MADDER
Irresistible to tiny moths

Sherardia arvensis | Rubiaceae

This charismatic little plant is considered by some to be a troublesome invasive weed – but what's not to like about its spiky whorls of bristly leaves and cross-shaped pale pink flowers? A native of dry grassland and arable fields, it turns up in lawns and flowerbeds but is just at home along the edges of pavements or cracks in tarmac. Its flowers are hermaphroditic and just 3–4 mm across. Attractive to a variety of insects, despite its unassuming appearance this species is important for invertebrates. The familiar **Cleavers *Galium aparine***, with its Velcro-stickiness and rounded burrs, is like a large climbing version of Field Madder.

DESCRIPTION: square, trailing stems, ringed by whorls of pointy dark green leaves, which can vary in length; sometimes forms a dense mat | **GROWTH FORM:** annual | **HEIGHT:** to 30 cm | **LEAVES:** whorls or four, five or six; young or stunted plants can have shorter, more oval-shaped leaves | **FLOWERING:** Apr–Oct | **FRUIT:** pairs of rough nutlets

MAUVE OR PURPLE FLOWERS WITH PROMINENT VEINS

LOVES THE EDGES OF PAVEMENTS AND VERGES

COMMON MALLOW
Resembles a cheese

Malva sylvestris | Malvaceae

Happy in places with plenty of nutrient enrichment, this species is readily encountered on verges and rough ground. Being hardy and drought tolerant, in some years it is particularly conspicuous when it comes into flower in late spring. As with Dwarf Mallow (see next species), the qualities of this plant led to it being used to make a soothing ointment and its fruits were sometimes eaten, hence the local name 'bread-and-cheese'. The dark pink-purple flowers are 3–4 cm across. Also look out for the beautiful **Musk Mallow *Malva moschata*** with its intricately divided leaves and showy pink or white flowers.

DESCRIPTION: purple-pink flowers with dark veins down the petals, stems erect to decumbent | **GROWTH FORM:** perennial | **HEIGHT:** 20–100 cm | **LEAVES:** dark green, three- or five-lobed | **FLOWERING:** ✿ May–Oct | **FRUIT:** flattened mericarps, each of which contains a nutlet

SMALL PALE PINK TO ALMOST WHITE FLOWERS

ROUNDISH, LOBED OR KIDNEY-SHAPED LEAVES

PROSTRATE TO ERECT

In the countryside, the plant likes to grow in nitrogen-rich locations such as stables and dung heaps. There are also places in urban areas where animal droppings provide plenty of nitrogen in the soil.

DWARF MALLOW
Resembles a little cheese

Malva neglecta | Malvaceae

The small, disc-shaped fruits (which have a pleasantly nutty flavour) look like small round wheels of cheese. Dwarf Mallow is native to Europe and western Asia and is widespread. It is an old medicinal plant that is traditionally used to treat sore throats and as an expectorant. It also contains vitamin C, saponins, tannins and mucilage. The pinky-white flowers are 1–2 cm across. This species can tolerate drier, more trampled areas than its larger cousin Common Mallow.

DESCRIPTION: pale pink to almost white flowers, vigorous growth, prostrate | **GROWTH FORM:** annual or biennial | **HEIGHT:** 10–50 cm | **LEAVES:** alternate; petiole up to 10 cm, leaves 2–6 cm long, roundish to reniform, divided into five to seven shallow lobes with serrate edges | **FLOWERING:** ❀ Jun–Oct | **FRUIT:** mericarps, each of which contains a nutlet

FIVE PINKISH-RED PETALS WITH TEN STAMENS

NEEDLE-LIKE LEAVES

PROSTRATE TO VERTICAL STEMS WITH SHORT HAIRS

SAND SPURREY
Delicate fuzzy beauty

Spergularia rubra | Caryophyllaceae

The tiny seeds are often dispersed by the wind, but frequently also by people, for example on their shoes or when soil is transported.

Blooming for less than a day, these delicate pink star-shaped flowers open only in fine weather, and close again in the afternoon. They can self-pollinate and are therefore not necessarily dependent on the flies that visit them. This graceful plant likes nutrient-rich, open locations and tolerates compacted soil, trampling and salt. It is therefore much more robust than it looks and copes well in the difficult urban environment.

DESCRIPTION: mostly prostrate, branching stems with clusters of axillary buds, silver-white gleaming stipules | **GROWTH FORM:** annual or herbaceous perennial | **HEIGHT:** 4–25 cm | **LEAVES:** opposite, entire, linear to thread-like, tapering to a fine point | **FLOWERING:** 🌸 May–Sep | **FRUIT:** capsule

LONG-STEMMED FLOWERS WITH FIVE STAMENS

PINNATE LEAVES

HAIRY, GENERALLY CREEPING STEMS

Common Storksbill has roots going down to depths of 1.5 m and can thus survive even in the dry cracks between paving stones. The spiral awn or tail on the seed can serve as a hygrometer to measure humidity.

COMMON STORKSBILL
Striking beak with special properties

Erodium cicutarium | Geraniaceae

Warm conditions are what the adaptable Common Storksbill really likes. It gets by with little water or nutrients and it tolerates trampling. This species does not need insects for pollination as it is self-pollinating. Its long seedheads, which look like a stork's beak, have an amazing property: in dry conditions, the lower section rolls up in a spiral. When it comes into contact with water, the awn stretches out again and burrows into the soil or into an animal's fur or moves along the ground as a creeper. Also look out for **Musky Storksbill** ***Erodium moschatum*** which has crinkly purple petals and less dissected leaves; it is swiftly becoming common in urban areas.

DESCRIPTION: flat, spreading leaf rosette | **GROWTH FORM:** annual or biennial | **HEIGHT:** 10–60 cm | **LEAVES:** deeply pinnately lobed to midrib with the pinnae themselves further divided | **FLOWERING:** 🌸 Apr–Oct | **FRUIT:** achene with long awn or tail

SEPALS PITCHER-SHAPED

FIVE-PART LEAVES, SOMETIMES HAND-SHAPED

GLANDULAR HAIRY

Grows both in shade and full sun. It has pigments that protect it from strong light and colour the plant dark red. The seeds can be propelled over distances of up to 6 metres.

HERB-ROBERT
Dainty flower with a strong smell

Geranium robertianum | Geraniaceae

This plant can cope with almost any location where sufficient nutrients are available. It can even climb up walls using its leaf petioles and side shoots. It is also known as 'Stinking Cranesbill' or 'Stinking Bob' because of its unpleasant, somewhat catty, scent. The story goes that Carl Linnaeus was on an excursion with his friend Robert and named the plant after him – because he was said to have a similar odour. An attractive white-flowered form sometimes occurs.

DESCRIPTION: delicate in appearance, often tinged with red and with lots of glandular hairs | **GROWTH FORM:** annual | **HEIGHT:** 20–40 cm | **LEAVES:** hand-shaped and divided into three or five leaflets, 3–4 cm long, 3–7.5 cm wide, can be directed towards the light by means of leaf articulations | **FLOWERING:** 🌸 May–Oct | **FRUIT:** capsule

BUNCHES OF DEEP PINK BLOOMS

THRIVES IN SUN-BAKED SPOTS

NARROWLY O... ENTIRE, OPPO... LEAVES

Despite their bitterness, the young leaves are eaten in France and Italy as a salad herb. This species is not useful for sleep: for that you need the roots of Common Valerian *Valeriana officinalis*, a plant of damp meadows and woodland edge.

RED VALERIAN
Manna for moths

Centranthus ruber | Valerianaceae

A native of southern Europe and the Mediterranean, this species has long been grown in gardens and flourishes as a very common escape, especially in towns, along railways and on coastal shingle. Its raspberry pink – sometimes white – flowers are not just admired by humans: they are much beloved of moths, particularly the day-flying Hummingbird Hawk-moth. If you are very lucky on a hot day in summer, you might just see this magnificent insect whizzing between the dense clusters of flowerheads, before hovering perfectly still as it drains the rich nectar from a bloom.

DESCRIPTION: erect, branched, hairless and grey-green | **GROWTH FORM:** perennial | **HEIGHT:** to 80 cm | **LEAVES:** ovate, opposite, somewhat fleshy; often curled along the edge when afflicted by aphid galls | **FLOWERING:** ❀ Apr–Oct | **FRUIT:** brownish and tapering, with fluffy pappus hairs

PURPLE DISC FLORETS IN A DENSE FLOWERHEAD

PRICKLY LEAVES

CREEPING THISTLE
Prickly friend to insects

Cirsium arvense | Asteraceae

The thistle, often unpopular due to its spikiness and association with desolate ground, is a firm favourite of insects. Numerous species of butterflies, including the Painted Lady, as well as many bees like to visit. The thistle gall fly *Urophora cardui* starts its life in the stem, while the thistlehead weevil *Rhinocyllus conicus* and another, rarer thistle weevil *Larinus turbinatus* are two attractive species that develop in the flowerheads. These plants react sensitively to fungal diseases and can be killed using organic means with a rust fungus.

> Incredibly vigorous: if the roots are hacked into bits, each tiny piece will soon develop into a new plant. A month after flowering, the new wind-borne seeds will already be germinating.

DESCRIPTION: leaves, not stems, are prickly; usually multiple 0.5–1 cm wide involucres | **GROWTH FORM:** herbaceous perennial | **HEIGHT:** 60–120 cm | **LEAVES:** alternate, undivided, prickly | **FLOWERING:** ✿ only disc florets, Jul–Sep | **FRUIT:** achenes with feathery hairs

YOUNG LEAVES REDDISH

PURPLE-LIPPED FLOWERS

SQUARE STEMS

The plant grows so quickly that it can produce four generations annually. Their flowers can even be seen in urban areas in winter if it is not too cold.

RED DEAD-NETTLE
Valuable to humans and animals

Lamium purpureum | Lamiaceae

The reddish colour of the young leaves affords good protection against intense sunlight. The flowers, which are already abundant as early as March, are very important for insects that are active in spring. The seeds are spread by ants, which eat the oily seed appendage. Few people are aware that the plant is also very healthy for us: both the leaves and the pretty flowers are edible, and they contain vitamins B and C and other beneficial compounds. You can eat them in a salad, prepare them like spinach or blend them into a smoothie. The moniker 'dead' derives from the fact that it does not sting.

DESCRIPTION: whole plant often with reddish or purple tinge, with a square stem | **GROWTH FORM:** annual or herbaceous perennial | **HEIGHT:** 15-45 cm | **LEAVES:** decussate (successive opposite leaf pairs 90 degrees apart), petiolate, 1-5 cm long, ovate to rounded ovate, irregularly notched or serrate, variably hairy | **FLOWERING:** ✿ Mar-Oct | **FRUIT:** groups of four nutlets

LONG BRIGHT MAGENTA FLOWERS
SPOT THE RED DEAD-NETTLE FOR COMPARISON
LEAVES IN A TIGHT RUFF-LIKE COLLAR

HENBIT DEAD-NETTLE

Long necks and frilly ruffs

Lamium amplexicaule | Lamiaceae

Yet another species that was largely known as a weed of field edges in Britain and Ireland until it began to decline in that habitat while at the same time becoming increasingly common in urban areas. Sometimes eaten as a salad herb. The flowers have an extended dark pink corolla. Probably a native of the Mediterranean, this species has successfully made its way around the world. Indeed, if you pay a visit to New York City you will find it to be common as a street weed there too!

DESCRIPTION: can look leggy, with long gaps between nodes, has a square stem | **GROWTH FORM:** annual | **HEIGHT:** to 25 cm | **LEAVES:** the pairs of toothed leaves are often fused and hug the stem, giving the appearance of a collar | **FLOWERING:** Mar–Nov | **FRUIT:** groups of four nutlets

CLUSTERS OF PALE PINK FLOWERS WITH DARK RED TIPS

DELICATE GREY-GREEN FOLIAGE

COMMON FUMITORY

Tiny ladders to nowhere in particular

Fumaria officinalis | Papaveraceae

This rather delightful little plant is generally considered an arable weed but it often appears on verges and as an unexpected bonus in municipal flowerbeds, as well as in allotments, gardens, old walls etc. The delicate, feathery blue-green leaves and the curious formation of the flowers are well worth close examination. Fumitory seeds can remain viable for a long time, so these plants may appear in a mass when soil is disturbed. The delicate appearance of the flowers and foliage gave rise to the genus name *Fumaria*, meaning 'smoke of the earth'.

DESCRIPTION: scrambling or upright plants with long racemes of flowers |
GROWTH FORM: annual | **HEIGHT:** to 15 cm | **LEAVES:** much divided and often
curving back | **FLOWERING:** ✖ Mar-Dec | **FRUIT:** rounded capsule

LEAF EDGES COARSELY TOOTHED

SKY-BLUE STRIPED FLOWER WITH YELLOW-WHITE CENTRE

Common Field-speedwell is green throughout the year and can, in contrast to our native speedwells, even flowers in winter if it is not too cold.

COMMON FIELD-SPEEDWELL

Flowers even in winter

Veronica persica | Plantaginaceae

In the early nineteenth century, this speedwell – introduced as an ornamental plant from western Asia – began to escape into the wild. Since then, it has been one of the most common speedwell species in Europe. It likes warm, open, nutrient-rich, loamy soil, where the roots grow down to depths of up to 20 cm. As with most species in the genus, Common Field-speedwell is pollinated by numerous small insects. The seeds have a nutritious appendage, so ants take the seeds into their nests as a food supply, where they germinate and grow.

DESCRIPTION: small sky-blue flowers with dark blue stripes, prostrate, hairy stems | **GROWTH FORM:** annual but may survive the winter | **HEIGHT:** 10–40 cm | **LEAVES:** alternate in lower part of stem, opposite in upper part, 1–2.5 cm in length, ovate, edge coarsely toothed | **FLOWERING:** 🍀 Jun–Aug | **FRUIT:** capsule

LEAVES MID-GREEN AND HAIRY

FLOWERS TINY AND FRAGILE, PETALS CAN BE DEEP BLUE ON UNDERSIDE

WALL SPEEDWELL
Tiny, bright and fragile

Veronica arvensis | Plantaginaceae

When growing on loose tarmac or on the top of walls these plants can be really minute, but the rich royal blue of the flowers – which are only about 3 mm across – will reward close scrutiny. Be careful though: as with some other speedwell species, even the slightest touch can dislodge the flowerhead so it will fall away. This herb has antiscorbutic and diuretic properties and has been used in traditional medicine to treat scrofula, among other things. Note: four or five other speedwells also occur in urban areas – if you find something different, take a look online where there are some good ID pages for the genus.

DESCRIPTION: tiny intense blue flowers arising from small upright plants | **GROWTH FORM:** annual, but may survive the winter | **HEIGHT:** to 20 cm | **LEAVES:** ovate, edges slightly to coarsely toothed; lower leaves with short stalks, upper leaves sessile | **FLOWERING:** ✿ Mar–Nov | **FRUIT:** two-lobed capsule

LONG CONICAL INFLORESCENCES ARE UNMISTAKABLE

NARROWLY OVATE, TONGUE-LIKE GREY-GREEN LEAVES

LISTEN FOR THE BUZZ OF BEES AND HOVERFLIES!

BUTTERFLY-BUSH
Cornucopia for proboscid friends

Buddleja davidii | Scrophulariaceae

The masses of small four-lobed flowers in long cones, usually some shade of purple but sometimes white, are a magnet for butterflies. This species originated in China and began to be cultivated in gardens at the end of the nineteenth century. It spread along railway lines initially, aided by its wind-dispersed seeds and with clinker as the ideal substrate. Often rather diminutive when growing on walls or in pavements, this plant can swiftly colonise wide areas on disturbed dry ground – to the detriment of native species – and so is considered to be invasive.

DESCRIPTION: vigorous and robust semi-evergreen shrub with grey-green foliage | **GROWTH FORM:** woody perennial | **HEIGHT:** up to 5 m | **LEAVES:** long, narrow, ovate and hairy, paler below | **FLOWERING:** ✤ Jun–Oct | **FRUIT:** capsules with winged seeds

FLOWERS POWDER BLUE WITH A YELLOW 'EYE' IN CENTRE

TONGUE-LIKE LEAVES

WHOLE PLANT IS HAIRY, ESPECIALLY SEPALS

FIELD FORGET-ME-NOT
Ol' blue eyes

Myosotis arvensis | Boraginaceae

It's always uplifting to see a forget-me-not: the bright powder-blue flowers with a yellow 'eye' at the centre. Legend has it that someone wearing a spray of these would not be forgotten by their lover. The plant is certainly hard for gardeners to forget, as it self-seeds readily and pops up all over the place. The flowers are 5 mm across (sometimes up to 7 mm). There are various cultivars and other species which have larger flowers, and pink-petalled morphs can turn up too. **Green Alkanet *Pentaglottis sempervirens*** is in the same family and has similar flowers but more of a sapphire blue with a white eye at the centre, and large rough leaves. It is very common on waste ground and on roadsides.

DESCRIPTION: hairy, ascending, branched, greyish-green, with spiralled stems of unmistakeable flowers | **GROWTH FORM:** annual | **HEIGHT:** 15–35 cm | **LEAVES:** oblong, hairy, with curved edges | **FLOWERING:** 🌸 Mar–June | **FRUIT:** small nutlets in the dried-up brown tubes

FLOWERS ARE REALLY TINY

PALE GREEN LEAVES

KEELED-FRUITED CORNSALAD
Smallest bunch of flowers you'll ever see

Valerianella carinata | Valerianaceae

The masses of minute flowers (just 1–2 mm across) have to be seen to be believed, but this plant soon becomes a familiar friend. The seeds germinate in the autumn and the plants flower early in the year, providing some welcome brightness on dim afternoons. Aided by milder winters, this species is becoming increasingly well established along pavements and walls as well as in gardens. **Common Cornsalad** *Valerianella locusta* is also frequent in similar habitats; it has a more flattened, rounded fruit with only a shallow groove.

DESCRIPTION: dense clusters of pale lilac flowers | **GROWTH FORM:** annual | **HEIGHT:** to 20 cm, often much less | **LEAVES:** spoon-shaped to oblong | **FLOWERING:** ❀ Feb–April | **FRUIT:** Egg-shaped fruits, rounded in section and longer than wide, with a distinct keel on one side

VIOLET BLUE FUNNEL-SHAPED FLOWERS

UPRIGHT STEMS

ROUNDISH KIDNEY-SHAPED LEAVES

This bellflower spreads mainly by seed, some of which are transported by animals. If a piece of the plant is detached, it can take root in another place and quickly form a broad cushion.

ADRIA BELLFLOWER

Escapee from rockeries and flowerpots

Campanula portenschlagiana | Campanulaceae

Popular with bees, this garden plant from Dalmatia, in the sunny south of Croatia, often escapes into the wild from rock gardens or containers. It is a cushion plant that can adapt to many different conditions, sunny or semi-shaded, dry or moist. It flowers several times a year and adorns pavements and driveways in our towns. It attracts honeybees and a wide variety of wild bees, including those that specialise in bellflowers. Also look out for **Trailing Bellflower *Campanula poscharskyana***, which is just as common on walls, which has larger, paler leaves and hairy flower buds.

DESCRIPTION: plant mealy grey-green, flower buds hairless | **GROWTH FORM:** herbaceous perennial | **HEIGHT:** 8–15 cm | **LEAVES:** alternate with roundish indented leaves | **FLOWERING:** 🌸 Jun–Aug | **FRUIT:** capsule

BLUE RAY FLORETS

STEMS ROUGH, BRANCHED AT TOP

LEAVES ON STEM ALTERNATE

WILD CHICORY

A favourite with the pantaloon bees

Cichorium intybus | Asteraceae

Widespread along paths and roads. Its deep roots means it is rarely found in pavement cracks. As a heat-loving and salt-tolerant plant, it can thrive in towns.

Chicory only blooms in the morning and its flowers are shut again by early afternoon. This is why it has a place on Carl Linnaeus' flower clock. Pantaloon bees (family Melittidae) have specialised in feeding from ray-floret composite plants such as Chicory; if no pantaloon bee or other bee species comes along to pollinate them, the flowers close a little later and the flower clock slows down.

The roots have been employed as a caffeine-free coffee substitute since the eighteenth century and this was commonly drunk during the Second World War when real coffee was scarce. Chicory is also used as a vegetable of course (and sometimes called 'endives' in such contexts).

DESCRIPTION: Chalky blue, dandelion-like flowers and leaves, but has taller, messy, branching growth; white sap | **GROWTH FORM:** herbaceous perennial | **HEIGHT:** 30–150 cm | **LEAVES:** alternate with dandelion-like leaves and stiff hairs on the underside | **FLOWERING:** ✿ only ray florets, Jul–Oct | **FRUIT:** achene

UPRIGHT SQUARE STEMS

LONGISH DEEPLY LOBED LEAVES

TINY VIOLET-WHITE OR PINK FLOWERS

VERVAIN
Traditional medicinal plant

Verbena officinalis | Verbenaceae

Grown in gardens since medieval times, Vervain is known as an old medicinal plant and was employed as an oxytocic: a drug that accelerates childbirth by stimulating uterine contractions. Related to Vervain is Lemon Verbena, the well-known culinary herb that smells strongly of lemon and, with its undivided leaves, looks very different. Vervain has a very long flowering season, often until October, and can be an important source of food for bees

The seeds of Vervain are spread by the wind. They also attach themselves to animals and shoes and thus can move rapidly through towns and villages.

late in the year. Also increasingly common on streets is the more recent garden escape **Argentinian Vervain** *Verbena bonariensis*, with is dense clusters of bright purple-pink flowers.

DESCRIPTION: very long, thin flower spikes with small violet-white or pink flowers, square stem | **GROWTH FORM:** annual herb | **HEIGHT:** 30-100 cm | **LEAVES:** opposite and deeply lobed | **FLOWERING:** ✳ Jul-Oct | **FRUIT:** schizocarp

PURPLE-BLUE, SLIGHTLY STRIPY LABIATE FLOWERS

SPRAWLING PLANT WITH ERECT FLOWERHEADS

LONGISH SESSILE LEAVES

'Reptans' means creeping and this is its strategy for success: its numerous runners take root when they touch the ground, thereby creating new plants.

BUGLE
Rambunctious purple carpet

Ajuga reptans | Lamiaceae

This native wild plant with its mauve-blue flowers is now deliberately planted as ground cover. It serves as a source of pollen for at least 11 species of wild bee and provides nectar and food for caterpillars of several species of butterfly. It is also a widely used medicinal plant in folk medicine: an infusion made from Bugle was said to help against rheumatism, stomach ulcers, angina and diarrhoea, while externally it is used for skin inflammations, haemorrhoids and catarrh (what ailment could it not heal?!). The leaves, stems and flowers are edible. They have a similarly bitter taste to Chicory and are suitable for flavouring or decorating.

DESCRIPTION: plant with basal, often reddish-tinged leaf rosette, long runners and bright blue, upright flowerheads | **GROWTH FORM:** herbaceous perennial | **HEIGHT:** 7–30 cm | **LEAVES:** decussate (opposite, but successive pairs at right angles to the pair below), tongue-shaped to spoon-shaped, entire or slightly indented, partly hairy | **FLOWERING:** 🌼 May–Aug | **FRUIT:** schizocarp

EVERGREEN LEAVES WITH STALKS

BLUISH-VIOLET FLOWER, TOOTHED WITH UPPER AND LOWER LIPS

ANGULAR STEM

Its habit of growing in gaps among other vegetation has helped it gain popularity. Gardening magazines recommend Selfheal as an attractive, evergreen space filler that draws insects to the patio or garden path.

SELFHEAL
Medicinal by name and by nature

Prunella vulgaris | Lamiaceae

The specific epithet of the scientific name, *prunella*, is derived from 'brunella', itself taken from 'Braune', brownness, the German name for diphtheria which the plant was once used to cure. Its ingredients have an antibacterial, healing and regenerative effect. In Asian medicine, this plant is used to treat fever, headaches and high blood pressure, for example. It is also used externally to help heal small wounds. The leaves taste slightly bitter and can be put in salads, while the flowers can be dried and used as a tea or in ointments.

DESCRIPTION: dense-flowered terminal heads on square stems, bright blue flowers with brownish tube | **GROWTH FORM:** herbaceous perennial | **HEIGHT:** 5–30 cm | **LEAVES:** opposite, with petiole, longish ovate, entire or slightly toothed | **FLOWERING:** Jun–Sep | **FRUIT:** capsule

RICH PURPLE FLOWERS

UPRIGHT, BRANCHING STEMS

LOTS OF SLENDER LEAVES

PURPLE TOADFLAX
A certain Italian flavour

Linaria purpurea | Plantaginaceae

The tall sprays of deep lilac (sometimes pink) flowers are as attractive to humans as they are to bees, hence this species has been grown in gardens for centuries and has been spreading outside them since at least the early 1800s. A familiar sight on waste ground, along railway lines and on the street, it is a close relative of Common Toadflax (see page 82) but unlike that native plant this one arrived on our shores from central and southern Italy. Vegetative shoots are noticeable throughout the winter months, when the leaves can look quite blue-grey.

DESCRIPTION: slim, upright plants with many flowers in a long raceme | **GROWTH FORM:** herbaceous perennial | **HEIGHT:** 30-90 cm | **LEAVES:** alternate, slender and strap-like with a pointed tip | **FLOWERING:** ✸ Jun-Oct | **FRUIT:** small capsule

FUNNEL-SHAPED FLOWERS
WITH UNEQUAL LOBES AND
PROTRUDING STAMENS

LEAVES ARE LINEAR-
LANCEOLATE

The resplendent Viper's-bugloss copes well in very dry, nutrient-poor soils, even those polluted with heavy metals. In urban areas it tends to grow at the edges of footpaths, pavements and on waste tips.

VIPER'S-BUGLOSS
Splendid urban elegance

Echium vulgare | Boraginaceae

This magnificent perennial is a real magnet for insects: it is visited by over 40 species of bees, bumblebees, hoverflies and butterflies. The colour of the flowers changes from red to blue as they lose their nectar, thus helping the insects to recognise which blooms are still worth visiting. In the first year, only the leaf rosette and the taproot are formed, and it is not until the second year that it flowers. Viper's-bugloss derives its name from its use in times gone by to treat snakebites.

DESCRIPTION: funnel-shaped flowers that spiral down the stem, initially pink, then blue; whole plant covered with bristly hairs | **GROWTH FORM:** herbaceous perennial | **HEIGHT:** 25–100 cm | **LEAVES:** evergreen, linear-lanceolate, alternate, up to 10 cm in length | **FLOWERING:** May–Sep | **FRUIT:** schizocarp

VIOLET FLOWERS
WITH SPUR

LEAF ROSETTE
WITH HEART-
SHAPED LEAVES

Unassuming groundcover with a bewitching scent, it has long been popular in gardens and parks and is now found everywhere in urban areas. Ants have contributed to this by distributing the seeds.

SWEET VIOLET
Welcome herald of spring

Viola odorata | Violaceae

As early as March, this plant delights us with its pretty, edible flowers. If you bend down sufficiently low you might be lucky enough to pick up its classic violet scent (sadly, not everyone can small this). Cherished as an ornamental and medicinal plant since the Middle Ages, this native species is widespread in Britain and Ireland, and is common in populated areas as well as lane sides and woodland edges. In addition to essential oils, which are used in perfumes, it also contains medically active substances that have a lipid-lowering and antibacterial effect. A white-flowered form is not uncommon.

DESCRIPTION: rosette with heart-shaped leaves, forms runners with new rosettes | **GROWTH FORM:** herbaceous perennial | **HEIGHT:** 5–15 cm | **LEAVES:** heart-shaped, only basal, undersides more or less glossy | **FLOWERING:** Mar–Apr | **FRUIT:** capsule

PALE VIOLET FLOWERS
WITH SPUR, YELLOW
INSIDE

REDDISH
STEMS

HEART-SHAPED,
LOBED LEAVES

This graceful summer-flowering plant with its pale violet-yellow flowers has a marked preference for the stone walls of old towns.

IVY-LEAVED TOADFLAX
Enchanting wall-illuminator

Cymbalaria muralis | Plantaginaceae

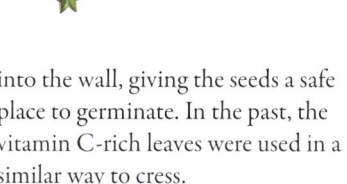

Originating in the Mediterranean region, this species grows on or near calcareous rocks and walls, especially if there are large moisture-filled cracks and crevices. A very clever little plant, it disperses some of its small seeds into the surrounding area whilst retaining one or two in the capsule. The plant then grows away from the light, back into the wall, giving the seeds a safe place to germinate. In the past, the vitamin C-rich leaves were used in a similar way to cress.

DESCRIPTION: pale violet flowers, borne singly on long reddish pedicels whose throat is closed by a pale yellow palate | **GROWTH FORM:** herbaceous perennial | **HEIGHT:** 10–40 cm, mostly prostrate or hanging | **LEAVES:** heart-shaped to round, underside red, leaf edges lobed | **FLOWERING:** ✿ Jun–Sep | **FRUIT:** capsule

ENTIRE PLANT COVERED WITH STINGING HAIRS

SERRATE POINTED LEAVES

ANGULAR STEM WITH OPPOSITE LEAVES AT RIGHT ANGLES TO THE PAIR BELOW (DECUSSATE)

A powerful rhizome forms runners, allowing the plant to reproduce vegetatively. The nutlets are ripe from September and are carried along by wind and rainwater or adhere to the coats of animals.

COMMON NETTLE
Nitrogen-loving superfood

Urtica dioica | Urticaceae

Nettles love nitrogen, which is why you can often find them in spots where dogs urinate. The Stinging Nettle frequently dominates the plant community in urban or rural areas where there is kitchen, garden or farm waste. Nettle seeds, with their nutty flavour, are a real superfood and contain potassium, iron, calcium and vitamins A, B and E. Not to mention nettle tea and nettle soup! This plant is an important foodplant for several species of butterfly too. Nettles need frost for germination and so may struggle in the warm, frost-free cities of the future.

Look out for **Small Nettle** *Urtica urens* too; a neater, prettier plant, its flowers are borne in tighter clusters and the whole thing looks more symmetrical.

DESCRIPTION: inflicts a burning sting when touched; angular stems | **GROWTH FORM:** perennial, erect plant | **HEIGHT:** 30–150 cm | **LEAVES:** decussate (successive opposite leaf pairs 90 degrees apart) with serrate, pointed leaves | **FLOWERING:** male flowers horizontal, female flowers hanging, Jun–Aug | **FRUIT:** nutlet

PLANTS OFTEN SCRUFFY LOOKING

CLUSTERS OF TINY FLOWERS IN LEAF AXILS

STEMS USUALLY REDDISH

PELLITORY-OF-THE-WALL
Friendly cousin of the Stinging Nettle

Parietaria judaica | Urticaceae

This plant lends a soft reddish tint to masonry and its dense groups of minute fluffy pinkish-green flowers repay close attention. The English name is tautologous, as 'pellitory' is derived from the Old French for 'of the wall', hence it really means 'wall-plant-of-the-wall'. But it does love walls!

Formerly employed to treat various ailments from kidney stones to burns, the species is most probably native here. It is also host to an exquisite little micromoth known as the Pellitory Beauty *Cosmopterix pulchrimella*, whose caterpillars make white 'windows' in the leaves as they feed.

DESCRIPTION: branching and ascending, with reddish somewhat hairy stems | **GROWTH FORM:** perennial | **HEIGHT:** to 30 cm, sometimes more | **LEAVES:** alternate, stalked, entire, oval, up to 5 cm long | **FLOWERING:** 🍀 Jun–Oct | **FRUIT:** small dark achenes clustered in the leaf axils

ROUNDISH FRUITS DIVIDED IN TWO

EXTENSIVELY BRANCHED GROWTH

NARROW-LEAVED PEPPERWORT

Unassuming and indestructible

Lepidium ruderale | Brassicaceae

The seeds, sometimes the whole plant, are torn off by the slipstream of cars or trains and dispersed along roads and railways.

From its original home, the dry, stony steppes of Eurasia, this little cress followed people into their settlements and cities. It can now be found almost everywhere in the world, and it does well in built-up areas. As the plant self-pollinates, it does not need coloured petals to attract insects.

Smells unpleasantly of urine and towards the end of the summer looks like a filigree candelabrum with multiple candles. The round, flattened seedpods are curiously charming.

DESCRIPTION: upright, very branched growth with clusters of dense flowerheads made up of minute florets | **GROWTH FORM** annual to biennial | **HEIGHT:** 10–30 cm | **LEAVES:** rosettes and lower leaves mostly pinnatifid, alternate, with petiole, upper leaves sessile, entire, linear to lanceolate | **FLOWERING:** 🍀 May–Oct | **FRUIT:** small pods

SWORD-LIKE LEAVES WITH FIVE PARALLEL LEAF VEINS

FLOWERHEAD WITH LONG STAMENS AND SHAPED LIKE AN EAR OF CORN

LEAF ROSETTE

Although Ribwort Plantain does not look as sturdy as Greater Plantain *Plantago major*, it copes very well in towns, thanks in part to its extensively branched, deeply penetrating roots.

RIBWORT PLANTAIN
Humble healing plant

Plantago lanceolata | Plantaginaceae

Like its cousin, Greater Plantain (see next species), Ribwort Plantain possesses many useful properties – it is an edible, medicinal herb which can tolerate heat and drought. However, it is not as resistant to trampling as 'Ironweed' and so prefers to grow where things are a little quieter. You can apply the juice from a crushed leaf directly onto small wounds or

insect bites – and this remedy is far more effective on nettle stings than the oft-cited dock leaves. There is one downside to the species, though: its pollen is a cause of hay fever.

DESCRIPTION: ovate-cylindrical flowerhead with long white-yellow stamens on a leafless furrowed stem | **GROWTH FORM:** herbaceous perennial | **HEIGHT:** 10–50 cm | **LEAVES:** basal rosette, narrow lanceolate, 10–20 cm long, up to 2 cm wide | **FLOWERING:** 🍀 May–Sep | **FRUIT:** capsule containing many seeds

SPOON-SHAPED LEAVES WITH FIVE TO NINE PARALLEL VEINS

YELLOW-GREEN-BROWN SPIKED FLOWERHEADS

LEAF ROSETTE

GREATER PLANTAIN
Frequently underfoot

Plantago major | Plantaginaceae

Dominates paths, as this is an extremely tough and adaptable plant. In good locations, it can form hand-sized, spoon-shaped leaves and almost knee-high flower stalks. On compacted soils or in narrow pavement cracks, in contrast, its leaf rosettes only grow a few centimetres in size, and it does not get very tall. It copes extremely well with trampling, salt and drought and obtains nutrients from its roots, which can reach down to 80 cm in depth – making it the perfect city dweller! European settlers brought the species to North America, where Indigenous peoples have called it the 'white man's footprint' ever since. Today, Greater Plantain – also know as Ironweed – can be found all over the world.

DESCRIPTION: spoon-shaped leaves with striking parallel veins, thin yellow-green-brown spike-shaped flowerheads somewhat resembling ears of corn | **GROWTH FORM:** herbaceous perennial | **HEIGHT:** 5–40 cm | **LEAVES:** basal rosette, over 10 cm long, on long petioles | **FLOWERING:** 🍀 Jun–Oct | **FRUIT:** capsules containing seeds

HEALING...

Greater plantain has been used in folk medicine for a long time and for many purposes. It contains numerous effective mucins (gel-forming compounds), bitter substances and tannins. For example, juice or broth made from the leaves has an anti-inflammatory effect and promotes the healing of wounds. Other healing effects, such as against viruses and ulcers, are still being researched. Plantain is considered a friend and protector of hikers as you can put the fresh leaves in your shoes as an insole, where they prevent or alleviate the formation of blisters.

Greater Plantain finds a place to grow even in the smallest crack in the middle of a road.

...AND EDIBLE

Plantain is rich in valuable vitamins and minerals. The young, soft leaves can be eaten raw in salads, prepared like spinach, consumed as a vegetable or added to soups. They can even be used to make a kind of sauerkraut. The small seeds contain a lot of protein and can be mixed into muesli, for instance, or ground into a healthy flour substitute. They store well when dried. You can also collect the young flower stalks and prepare them like asparagus – they taste similar to mushrooms. The long taproots offer a vegetable with an intense flavour.

When they get moist, the casings of the ripe seeds swell and become sticky and slimy. They thus adhere to animal paws, shoes and tyres and in this way have been spread all over the world.

ALTERNATE LIGHT GREEN LEAVES

FLOWERS IN GREEN FALSE UMBELS

MILKY SAP OOZES OUT OF THE STEM IF IT IS DAMAGED

PETTY SPURGE
Delicate citizen of the world

Euphorbia peplus | Euphorbiaceae

The white sap that leaks out when you touch this plant is poisonous and can cause skin irritation. The flowers consist only of the reproductive organs, which are surrounded by yellow-green bracts. This entire structure is actually called a 'false flower' or pseudanthium. The nectar glands offer the nectar openly as a sparkling drop of liquid – very inviting for insects which can simply lap it up. When ripe, the seed capsules explode and project the seeds effectively into the surrounding area.

Originating from Southern Europe and North Africa, Petty Spurge has spread worldwide. Its great adaptability helps it to colonise urban areas.

DESCRIPTION: yellow-green false flowers with four oval, yellow glands | **GROWTH FORM:** annual | **HEIGHT:** 10–30 cm | **LEAVES:** roundish to obovate with blunt tip, entire, 0.5–2 cm | **FLOWERING:** 🍀 Jul-Oct | **FRUIT:** capsule

GREENISH
FLOWER SPIKES

OPPOSITE AND
SERRATE EDGES

MALE AND FEMALE
FLOWERS ON
DIFFERENT PLANTS

DOES NOT HAVE
MILKY SAP

ANNUAL MERCURY

Distinct male and female plants

Mercurialis annua | Euphorbiaceae

Male and female flowers are usually found on different plants, a phenomenon known as dioecy. Annual Mercury is an ancient medicinal plant and is considered to be a diuretic, a laxative and effective for women's ailments. The scientific genus name *Mercurialis* goes back to Mercury, the Roman messenger of the gods who,

Originally from Southern Europe, this plant likes warm, nutrient-rich locations where its roots reach down to 50 cm in depth. It is perfect for the city and can project its seeds up to 4 metres away.

according to legend, discovered the healing powers of this species. A single plant can produce over one billion pollen grains – no wonder it is such a common weed of gardens and waste ground.

DESCRIPTION: alternate branched plant with long, greenish flower spikes | **GROWTH FORM:** herbaceous perennial | **HEIGHT:** 20–50 cm | **LEAVES:** opposite, bare, narrow ovate to lanceolate and on the edge bluntly toothed, 3–10 cm long | **FLOWERING:** 🍀 May-Oct | **FRUIT:** capsule with single-seeded mericarps

SPIKE OF GREYISH FLOWERS IN KNOTTED CLUSTERS

MEALY GREY-GREEN PLANT

ALTERNATE SERRATE LEAVES

FAT-HEN
Food for butterflies, transmits viruses

Chenopodium album | Amaranthaceae

Widespread in agricultural root crops, vegetable gardens and also frequent in urban areas, you can find this pioneer plant on rubble heaps, building sites and by paths. In our country childhood, we would be constantly pulling it out of the ground between root crops such as turnips and potatoes. Fat-hen does not tolerate salt and dries out on contact with it.

AN OLD AGRICULTURAL CROP, its seeds were being used to make flour as far back as the Neolithic. The leaves are still prepared today like spinach or used as an additive in smoothies and salads because they contain many vitamins and minerals. Do be careful, though, as this plant contains relatively high levels of oxalic acid (as do certain other edible plants such as parsley) and should therefore not be

DESCRIPTION: Mealy grey-green plant | **GROWTH FORM:** annual | **HEIGHT:** 20–150 cm | **LEAVES:** alternate with oval-lanceolate or diamond-shaped leaves | **FLOWERING:** Jul–Oct | **FRUIT:** single nutlet

The seeds can lie buried for several years before germination. It spreads rapidly on new bare soil because it is eaten but not digested by most animals.

eaten in large quantities. Many similar species exist and can turn up occasionally in towns and port areas.

IT HOSTS MANY PESTS such as the black bean aphid and a whole variety of plant viruses. However, it also provides a habitat for many butterfly and moth caterpillars worldwide, such as skippers. In the scientific literature, we have found over 70 species that feed on Fat-hen. However, it is not attractive to flower-visiting insects as it is adapted to wind pollination and does not produce nectar. The similar species **Fig-leaved Goosefoot** *Chenopodium ficifolium* can also turn up; its lower leaves are distinctly three lobed, while upper leaves can be narrowly ovate.

The inconspicuous flowers of Fat-hen are pollinated by the wind.

Fat-hen is the foodplant for the caterpillars of over 70 butterfly and moth species.

EVERGREEN WITH ADHESIVE ROOTS

NON-FLOWERING SHOOTS WITH THREE- TO FIVE-LOBED LEAVES

IVY
A real social climber

Hedera helix | Araliaceae

Ivy leaves have been employed as a painkiller since ancient times. Today they are often used to treat respiratory infections. However, the juice of the berries and the fresh leaves may trigger allergic reactions on the skin or cause nausea, diarrhoea and vomiting if consumed.

A LOT OF MEANING has always been attached to Ivy. Its evergreen leaves symbolise loyalty, reliability and eternal life. Ivy also has a symbolic connection to the deities of ancient Egypt and Greece, and in some wine-growing regions, the appearance of Ivy fruit in spring is believed to indicate the success or otherwise of the grape harvest. In Britain it was held that Ivy kept evil away from the house or one's livestock. Not a lot of people know that there are actually two native species of ivy in Britain and Ireland: Atlantic Ivy *Hedera hibernica*, which is widespread in the west, can be differentiated by its more pleasant resinous

DESCRIPTION: evergreen with holdfast roots, round umbels | **GROWTH FORM:** climber | **HEIGHT:** up to 20 m | **LEAVES:** alternate, three to five lobes for the non-flowering shoots, ovate to diamond-shaped for the flowering ones | **FLOWERING:** Sep–Nov | **FRUIT:** blue-black berries arranged in an umbel

smell and by having hairs which lie flat, rather than being spiky.

SPECIAL FOOD SOURCE This native climber of trees and rock, which can live for up to 250 years, provides food for a great many bees, wasps, flies, butterflies and beetles during to its autumn flowering period. It even has its own attractive species of bee, the Ivy Bee *Colletes hederae*, which has a particular fondness for ivy pollen. But that's not all, because its berries also ripen at an unusual time, between January and April, thus offering many birds a juicy meal in the middle of winter. You can often see the likes of blackbirds, redwings, starlings and robins feeding on the fruits of ivy.

NATURAL THERMAL INSULATION can be provided by Ivy growth on buildings (despite much hearsay, it is scarcely damaging to masonry). Ivy gives life to walls and fences by providing insects and birds with food and nesting places in a wall–ivy symbiosis, at the same time protecting the wall or fence from weathering. It also has an insulating effect, especially against summer heat. With its adhesive roots, it does not shy away from any wall or façade and climbs up wherever it can.

Ivy has its own species of bee, the Ivy Bee *Colletes hederae*.

Food, nesting or hiding places: habitat and natural façade cover – ivy is multi-functional.

A very tenacious plant, Ivy creeps along the ground and takes over any non-concrete corners, stretching out its holdfast roots and climbing up house fronts and trees.

YOUNG LEAVES ARE TINGED ORANGE-RED

ALTERNATE, PINNATE LEAVES WITH A SINGLE LEAFLET AT THE TIP

NO STIPULES

TREE-OF-HEAVEN
Sucker for spreading fast

Ailanthus altissima | Simaroubaceae

> Being salt-tolerant, resistant to herbicides and requiring a lot of light, this species is perfectly primed to thrive in urban areas.

This species is nowadays considered to be the fastest-growing tree in Europe. After germination, the seedlings can grow to a height of 1 m in the same year, and even up to 3 m if they find themselves in deep, fresh and nutrient-rich soil. As well as seed, spreads very effectively by suckering. In its original homeland of China, the leaves serve as food for the magnificent Ailanthus Silkmoth *Samia cynthia*. Not so elsewhere – indeed, in many places this species has become so invasive that it is perhaps more popularly known as 'Tree-of-Hell'.

DESCRIPTION: flowers smell strong and unpleasant and are pollinated by honeybees, beetles and flies | **GROWTH FORM:** | **HEIGHT:** up to 25 m | **LEAVES:** alternate, pinnate with pointed and partially lobed leaflets | **FLOWERING:** greenish flowers in panicles in summer | **FRUIT:** winged fruit, similar to Ash *Fraxinus excelsior*

FIVE-LOBED LEAVES

SERRATE LEAF EDGE

The winged seeds of Sycamore and other maple species can be carried for several kilometres on stormy days. Sycamore appreciates sunlight and therefore often colonises urban areas.

SYCAMORE
Insect playground

Acer pseudoplatanus | Sapindaceae

Its seedlings grow on walls and waysides, but they often lack the space to become a flowering tree. A native of Central and Southern Europe, this species has long been naturalised on our shores. The leaves provide habitat and food for beetles, bees, wasps, bugs, mites and moths in the urban environment. Indeed, close inspection of the underside of leaves may reveal millions of aphids drinking the sap – which can be a nuisance for car owners if they park beneath and their vehicle ends up covered in sticky honeydew (aphid poo)! It is wonderful when the Sycamore is allowed to flower, however, as its large quantities of nectar provides many bees with valuable carbohydrates.

DESCRIPTION: its seeds – contained in winged samaras – spin rapidly when airborne, aiding dispersal | **GROWTH FORM:** 🌳 | **HEIGHT:** up to 25 m | **LEAVES:** opposite, five-lobed, irregularly serrate leaf edge | **FLOWERING:** May | **FRUIT:** schizocarp with samaras in a hanging cluster

ALL PARTS OF THE PLANT OFTEN PURPLE-TINGED

STALKS INITIALLY PROSTRATE AND THEN ASCENDING VERTICALLY

HAIRY FINGER-GRASS
Tough little cereal

Digitaria sanguinalis | Poaceae

Originating in southeastern Europe, this grass is now commonly found worldwide in the cracks of paving stones. Its roots extend down to depths of 35 cm and its tiny seeds are often spread by ants.

Nowadays this species is usually trampled underfoot, but in the past it was widely cultivated as a cereal and its seeds were used until the nineteenth century to cook nutritious porridges. In the USA, it is still cultivated as forage. In agriculture this species is considered an undesirable weed and keen gardeners will discuss in detail how best to eradicate it from their lawns. It appreciates dry, nutrient-rich soil in warm locations and is considered a beneficiary of climate change.

DESCRIPTION: frequently blood-red colour to the parts of the plant above the soil, prostrate flat rosette, then ascending | **GROWTH FORM:** annual | **HEIGHT:** 10–60 cm | **LEAVES:** flat, 3–10 cm long, up to 8 mm wide, pointed tip | **FLOWERING:** Aug–Oct | **FRUIT:** grain up to 4 mm long

DISTINCTIVE PALE-TIPPED SPIKELETS

PEA-GREEN LEAF BLADES

WATER BENT
The ultimate superspreader

Polypogon viridis | Poaceae

The 'bent' grasses, an informal grouping covering more than one genus, are so named because their leaf blades stick out at a pronounced angle. In North America, this particular species is known as Beardless Rabbit's-foot Grass.

This native of dampish habitats in the Mediterranean has been known from the Channel Islands and a few places in southern England since the mid-1800s. It proceeded to spread gradually, before a sudden and very rapid surge in distribution commenced in the latter years of the twentieth century. It is now ubiquitous in many areas, having undergone one of the most dramatic range expansions of any alien plant. As with many such species, human activities (such as building work and horticulture) and climate change clearly have a major part to play.

DESCRIPTION: flowering spikes that go from fluffy to scruffy, spreads by stolons which root at the nodes as well as by seed | **GROWTH FORM:** annual or perennial | **HEIGHT:** 10–60 cm | **LEAVES:** hairless, slightly curved on the back, 3–18 cm long, pointed tip | **FLOWERING:** May–Oct | **FRUIT:** pale brown grain about 1 mm long

SIMILAR TO A BARLEY SPIKE

LEAVES EXTENDING UP TO JUST UNDER THE INFLORESCENCE

LEAF SHEATH OF THE UPPERMOST LEAF INFLATED

Barbs on the awn enable the seeds of Wall Barley to hook onto the fur of dogs and cats as well as squirrels and other animals, ensuring a good dispersal.

WALL BARLEY
Irritates cats and dogs

Hordeum murinum | Poaceae

This wild grain found in villages and towns is actually quite tasty. However, as large mammals, we have too little patience to collect the tiny seeds, in contrast to mice! Furthermore, Wall Barley is also known as 'dog-wee grass' thanks to its penchant for growing in the nitrogen-rich soil caused by dogs leaving scent marks. The awns have barbs which can irritate and injure the mucous membranes of dogs and cats: these barbs migrate into the ears, noses and eyes of the animals and may also cause severe inflammation under the skin.

DESCRIPTION: closely resembles standard agricultural barley with its long awns, except it is much smaller | **GROWTH FORM:** | **HEIGHT:** 10–40 cm | **LEAVES:** alternate, simple lanceolate, sheaths rounded on the back | **FLOWERING:** May–Oct | **FRUIT:** grain up to about 6 mm long

SPIKELETS LIE WITH THEIR LONG SIDE NEAREST THE STALK

STALK SQUIGGLY ZIG-ZAGS AT THE INFLORESCENCE

STALKS UPWARD CURVING

Popular fodder and lawn grass. When found in town, it is usually a garden escape. This grass reproduces readily and quickly by means of runners.

PERENNIAL RYE-GRASS
Shows great resilience

Lolium perenne | Poaceae

The specific epithet *perenne* means 'everlasting'. This grass grows very quickly, tolerates trampling and frequent cutting and is therefore an important component of ornamental and commercial lawns. At the same time, the species is a very good and productive fodder grass for pasture. Because of these properties, it has spread from Europe throughout the whole world. Ryegrass strongly favours well-fertilised soils and in the wild is found mainly in nutrient-rich meadows. It has a typical 'shiny' look when seen in a mass.

DESCRIPTION: flat spikelets up to 20 mm long, with their narrow side attached to the stalk | **GROWTH FORM:** | **HEIGHT:** 10–70 cm | **LEAVES:** dark green shiny leaves, 2–4 mm wide and up to 20 cm long | **FLOWERING:** May–Oct | **FRUIT:** grain up to about 4 mm long, in hardened casing

LEAVES BLUNT, TIP SHAPED LIKE THE PROW OF A CANOE

FORMS SMALL CLUSTERS

This short-growing grass may even flower and bear seed in winter. Seeds often disperse both in the warm and cold seasons on our shoes, finding a home in a nearby crack in the pavement.

ANNUAL MEADOW-GRASS
Robust and ever-present

Poa annua | Poaceae

Characteristic grass of areas subject to heavy trampling, it grows worldwide from the Tropics to the Antarctic in gaps and pavement cracks, in gardens and fields. Indeed, there is a strong case to be made that this is the most widespread plant species in the world – and so it surely deserves our admiration, especially as it is so unassuming. As a nitrogen-loving species, we can also find Annual Meadow-grass in places where dogs and other animals regularly urinate. This shallow-rooted plant is unpopular in gardens as it turns yellow too quickly and is also susceptible to fungal diseases.

DESCRIPTION: short grass with panicles throughout the year | **GROWTH FORM:** in clusters | **HEIGHT:** 2–30 cm | **LEAVES:** alternate, linear, prow-shaped tip | **FLOWERING:** Mar–Dec | **FRUIT:** seed with stiff-haired bracts

LONG DROOPING INFLORESCENCES

OFTEN SHOWS PURPLISH-RED TONES AS WELL

EXTENDED AWNS ON THE SPIKELETS

BARREN BROME
Drooping and dangling

Anisantha sterilis | Poaceae

A rather elegant grass; green or purplish, later pale buff-brown in colour, with long awns. A crop weed of farmland, this species is just as at home on pavements and in edgelands of all sorts, where it may occur in substantial swathes. Several other similar brome grasses sometimes appear in such situations, but most of these have more congested inflorescences. One or two of the florets near the tip are usually sterile, which explains the name. Considered an aggressive invasive weed in North America.

DESCRIPTION: arching grass with nodding panicles of widely spaced spikelets on long wiry stalks | **GROWTH FORM:** annual or biennial | **HEIGHT:** 15–100 cm | **LEAVES:** softly hairy, up to 25 cm long | **FLOWERING:** Apr–Jul | **FRUIT:** long thin grain, hairy at the tip

UPRIGHT STEMS
WITH DELICATE
DANGLING
INFLORESCENCES

SPIKELET COVERED
IN TINY HAIRS

FLORETS DENSELY
PACKED WITHIN
EACH SPIKELET

GREATER QUAKING-GRASS

Can you hear a rattling?

Briza maxima | Poaceae

This striking grass is popular as an ornamental in gardens and often escapes onto pavements and proceeds to spread apace. The large dangling silvery-green spikelets with their papery lemmas bear a passing resemblance to the tip of a rattlesnake's tail, but are rather more welcomingly tactile! A Mediterranean species, it is now widely naturalised in many places around the world and is popular as a dried specimen in wreaths and floral displays.

DESCRIPTION: upright grass with large overlapping inflorescences on dangling pedicels | **GROWTH FORM:** annual | **HEIGHT:** 25–70 cm | **LEAVES:** smooth and hairless, often purple tinged | **FLOWERING:** Apr–Jul | **FRUIT:** an almost circular grain about 2.5 mm across

YELLOW-GREEN TO RED-BROWN CLUSTERS OF SEED CAPSULES

SHARP POINTS AT THE END OF THE LEAF SHEATHS

STIFF STEMS WITHOUT NODES

SLENDER RUSH
Tough globetrotter

Juncus tenuis | Juncaceae

This species, which is more usually associated with watery habitats, can be found on footpaths, roadsides and in holes in paving stones. It is very adaptable and thrives in full sun, in dry or moist conditions. It can also colonise bare, compacted soil as a pioneer plant. Originally from North America, it has spread to Europe since the nineteenth century and can now be encountered all over the world. Indeed, it has expanded its range across Britain and Ireland in recent decades. Its fibres were once used to make string.

Moisten a ripe capsule (seed) of a rush with some water. After a few minutes the swollen seeds emerge resembling frog spawn. They can still germinate even after long periods of time.

DESCRIPTION: small yellow-green to red-brown tussocks with terminal inflorescences | **GROWTH FORM:** perennial | **HEIGHT:** 15–50 cm | **LEAVES:** peduncles have one or two grass-like leaves or bracts | **FLOWERING:** Jun–Sep | **FRUIT:** capsule

VERY ROBUST PLANTS

HORIZONTAL, SPREADING FORM WHEN GROWING ON GRAVEL OR TARMAC

BRANCHES IN WHORLS AROUND THE STEM

Early in the year, this species puts up fertile stems with brownish spore-bearing cones. These look almost like strange fungi.

FIELD HORSETAIL
Good for buffing pewter

Equisetum arvense | Equisetaceae

Able to spread through rhizome fragments as well as spores, and strongly resistant to herbicides, this species can be an unwelcome guest in gardens and allotments. Often appears in stonework and pavements, especially in its brush-like sprawling form (see inset above). Due to having a high silica content, the plant can be employed as an abrasive and so in the past was used for polishing. It contains fungicidal compounds that can help combat crop disease. Also look out for **Great Horsetail** *Equisetum telmateia*, which has thick whitish main stems and gets very big, almost to 2 m tall.

DESCRIPTION: ridged, spindly and deciduous; upright or prostrate | **GROWTH FORM:** smooth stems with many crowded whorls of branchlets | **HEIGHT:** to 80 cm | **LEAVES:** simple pinnate frond consisting of many roundish, serrate leaflets | **FRUIT:** spore cones at the end of special unbranched brown stems, which appear early in the year, before the vegetative stems

NEW LEAVES PALE GREEN BECOMING DARKER WITH AGE

LEAVES OFTEN WITH WAVY EDGES BUT ALL ONE PIECE

LIKE MANY URBAN PLANTS, THIS FERN CAN BE CAUGHT IN THE FIRING LINE OF GRAFFITI ART

HART'S-TONGUE FERN
A forest spirit in the city

Asplenium scolopendrium | Aspleniaceae

Unmistakeable with its bright green undivided fronds. These can have a wavy edge, but are always entire. This is a plant of shady walls and damp corners, often growing intermingled with bryophytes and some of the other ferns listed in this section; when on brickwork and the like it can be very small. Its name comes from the resemblance to the tongue of a male red deer, for which 'hart' is an antiquated term. Long popular as a garden plant, there are many cultivated varieties with different levels of crinkliness.

DESCRIPTION: pale frond stalk, evergreen | **GROWTH FORM:** single or multiple fronds | **LENGTH:** 5–50 cm | **LEAVES:** simple, non-pinnate, entire smooth frond | **FRUIT:** spore capsules in stripes on the underside of the leaf, Jul–Sept

NEATLY ROUNDED PINNAE

FRONDS VERY DIFFERENT TOP AND BOTTOM

SHRIVELS UP IN DRY CONDITIONS

RUSTYBACK
Curls up in the heat

This species holds some secrets: it contains potent compounds and enzymes to aid recovery after drying out. It has also been shown to have anti-cancer and antibacterial properties.

Asplenium ceterach | Aspleniaceae

The pinnate fronds of this winsome little fern have rounded lobes and a distinctive two-tone appearance, with pale-green uppersides and undersides covered in thick whitish scales which gradually fade to bronzy brown – hence the name. A calcicole, it favours limestone and other calcareous rocky substrates and so is common on walls in some areas. With a marked ability to withstand drying out, the fronds curl up as they wait for the rains to return. Perhaps the most tactile and irresistible of our fern species.

DESCRIPTION: furry undersides, evergreen | **GROWTH FORM:** dense clusters | **LENGTH:** 3–15 cm | **LEAVES:** simple pinnate frond with rounded pinnae | **FRUIT:** prolific spore capsules on underside of the leaf, Jul–Oct

ROUNDISH SERRATE LEAVES
ON BROWN PETIOLE

(SPOT THE CAMEO
FROM THE NEXT
SPECIES)

SIMPLE PINNATE
FROND

This inhabitant of dry walls shows just how undemanding and adaptable a fern can be that actually prefers moist conditions. And how pretty too!

MAIDENHAIR SPLEENWORT
Leads a shadowy existence

Asplenium trichomanes | Aspleniaceae

The brown-stemmed Maidenhair Spleenwort loves to dwell in damp and shady cracks in urban walls. It is particularly fond of wells, streams and other water sources in villages and towns. However, it also likes shady driveways and sometimes grows there together with liverworts such as Common Liverwort *Marchantia polymorpha*. This species is actually becoming increasingly common in cities, as it continues to evolve and adapt to modern life.

DESCRIPTION: brown stems, evergreen | **GROWTH FORM:** clusters | **HEIGHT:** 5–25 cm | **LEAVES:** simple pinnate frond consisting of many roundish, minutely serrate leaflets | **FRUIT:** spore capsules over whole underside of the leaf, Jul–Aug

BIPINNATE OR TRIPINNATE FRONDS

DIAMOND-SHAPED PINNAE

Thanks to its thickened rhizome, Wall-rue is able to store water and survive when the cracks in the wall are dry. Also occasionally grows in paving stone cracks.

WALL-RUE
Evolution in your area

Asplenium ruta-muraria | Aspleniaceae

A characteristic species of a plant community that specialises in cracks in walls and mortar joints. Originally, of course, it only occurred in rock crevices in mountains. Acid rain had a negative effect on this species, but it has come back strongly since sulphur dioxide levels have dropped. This small fern is highly variable in its appearance, and it is likely that due to the intricacies of its evolution what we are actually looking at is a complex of many cryptic species.

DESCRIPTION: diamond-shaped pinnae, evergreen | **GROWTH FORM:** | **HEIGHT:** 10–30 cm | **LEAVES:** bipinnate or tripinnate fronds with irregular triangular to oval leaflets | **FRUIT:** spore capsules over the entire underside of the leaf, Jul–Oct

LEAVES TIGHTLY OVERLAP, MAKING SHOOTS LOOK WORM-LIKE

FORMS DENSE CUSHIONS OF CLOSELY PACKED SHOOTS

WHITISH TIPS TO THE LEAVES GIVE A SILVERY APPEARANCE

SILVER-MOSS
Shimmer from the gutter

Bryum argenteum | Bryaceae

Try viewing this moss from different angles to observe its characteristic metallic sheen. It often grows with other mosses too – how do they differ?

The most common moss in urban places, it even grows in the cracks of roofs or window frames. You can find Silver-moss almost everywhere in the centre of large cities, often where no other plants grow. It is perfectly adapted to drought, as the outer cells die off and form a protective layer around the rest of the tissue. This and its preference for high nitrogen levels and tolerance of air pollution mean that it has been able to colonise cracks in pavements all over the world. Its importance for the ecosystem should not be underestimated, as it offers a home to microscopic life in some of the harshest environments.

DESCRIPTION: silvery sheen with pear-shaped, hanging capsules | **GROWTH FORM:** cushions or stray shoots | **HEIGHT:** 0.2–2 cm | **LEAVES:** spiral-shaped, scaly, pointed | **FRUIT:** hanging dark to red-brown capsules

GREYISH GLASS-LIKE HAIRS ON THE LEAF TIPS

SPHERICAL OR HEMISPHERICAL CUSHION

CAPSULE WITH EIGHT TO TEN LONGITUDINAL RIBS

Dust, earth, sand and other fine-grained materials get stuck to this unassuming moss. In this way soil is formed that can then be colonised by flowering plants in the urban environment.

GREY-CUSHIONED GRIMMIA
In a fluffy grey coat

Grimmia pulvinata | Grimmiaceae

Hemispherical cushions found on almost any roof or wall. The grey colouring that shrouds the moss like a coat is formed by long white hairs on the tip of each leaflet, an adaptation to drought and heat. This species is sensitive to salt and thrives without nutrients. Like all mosses, it stores water and only releases it slowly in dry conditions. Moss cushions are therefore important for the urban climate as regulators of the water regime.

There are many other species of bryophytes (mosses and liverworts) that appear in urban places. The British Bryological Society has great resources to help beginners, e.g.: https://www.britishbryologicalsociety.org.uk/learning/some-common-bryophytes/

DESCRIPTION: spherical or hemispherical cushion | **GROWTH FORM:** cushion moss | **HEIGHT:** 0.5–3 cm | **LEAVES:** lanceolate with hair tip | **FRUIT:** hanging capsules initially inclining towards the moss cushion before eventually becoming erect

STEMS OF SPOROPHYTES

GEMMAE CUPS

CREEPING GREEN MATS

COMMON LIVERWORT
Kerb crawler par excellence

Marchantia polymorpha | Marchantiaceae

Alongside their cousins the mosses and hornworts, liverworts are part of the ancient lineage of plants known as bryophytes. Like ferns they do not flower, but reproduce by spores – as well as other means, including gemmae (tiny clones). This species is very common in urban areas, where it loves pavement cracks and the bottoms of walls. A thalloid liverwort – i.e. it spreads in a mat (thallus), rather than having tiny leaves – it produces little umbrella-like sporophytes and also sports minute cups on the surface, which are full of gemmae. When it rains, the gemmae are splashed out of the cups to make new plants nearby.

DESCRIPTION: dark green, crinkly edges, often with a dark midrib | **GROWTH FORM:** | **HEIGHT:** flat to the ground | **LEAVES:** a spreading thallus that bifurcates again and again | **FRUIT:** spores on upright structures and gemmae in surface cups

Glossary

We have tried to use as few specialist terms as possible, although sometimes they are unavoidable. If you come across unfamiliar words and expressions, this is the place to find out what they mean.

achene: a simple fruit structure, containing a single seed

adhesive root: shoot-like roots on climbing plants that allow them to anchor themselves in cracks or crevices

alternate: occurring singly in different places along the stem (as opposed to opposite)

annual: completes lifecycle in a single year or growing season

anther: the pollen-bearing tip of a stamen, the male parts of a flower

awn: a bristle-like projection on grasses, often long

axillary: occurring in an axil, the area between a leaf and the main stem

basal: leaves at the bottom of a main stem, usually on or near the ground

biennial: lifecycle takes two years or growing seasons to complete

bipinnate: dividing and then dividing again

bract: small leaf-like structure, usually in the upper stem or around a flowerhead

burr: a rough or prickly fruit designed to be carried accidentally by animals

capsule: spore container (sporangium) found in mosses

carpel: female reproductive organ of a flower (later it forms the fruit)

congener: a species in the same genus

cushion: compact growth form in perennials, typical of desert, rock and high-mountain plants; also exhibited by many mosses

decumbent: with branches or stems that grow along the ground but then turn upwards towards the end

decussate: leaves in opposite pairs, with each pair at a right-angle to the next

dehiscent: breaking up at maturity

dioecious: plants that have only male or only female flowers on any individual plant

disc floret: minute flower with no projection, comprising part of a capitulum (flowerhead)

entire: without teeth or lobes, having smooth edge

follicle: pod-like fruit containing one or more seeds

frond: leaf-like structure of a fern

genus: a group of closely related species within the same family

glandular: tipped with tiny globules of sticky, sometimes aromatic liquid

involucre: dense ring of bracts below an inflorescence

lanceolate: elongated, widest below middle and tapering to tip

linear: many times longer than wide, strap-shaped

lobe: a clear part of single divided leaf, can be shallow or deep

mericarp: individual portion of a schizocarp, each one containing a single seed

midrib: central, main vein of a leaf

monoecious: plants that have male and female flowers on the same individual

mycorrhiza: community between a plant and a fungus on or in the roots; the fungus benefits from sugar from the plant and in turn the plant itself derives nutrients from the fungus

neophyte: a plant that establishes itself in areas where it is not native

node: part of a stem from which leaves or branches arise

oblanceolate: elongated, widest above middle and tapering to base (inverse of lanceolate)

obovate: widest above middle and tapering to base (inverse of ovate)

odd pinnate: having an unequal number of leaflets, hence a single terminal leaflet is present

opposite: leaves borne at the same level, on opposite sides of the stem

ovate: widest below middle and tapering to tip, egg-shaped

panicle: inflorescence where individual flowers are borne on branches off the main stem (common in grasses)

pappus: feathery hairs attached to an achene, helping it float on the breeze (think dandelion clocks)

pedicel: stalk of single flower

perennial: persists for at least several years

petiolate: with a leaf stalk

petiole: leaf stalk

phyllary: one of the involucral bracts around a composite flowerhead

pinnae (singular: pinna): segments of a compound leaf

pinnate: compound leaf, with pinnae or leaflets arranged as opposite pairs along a single stem

pinnatifid: pinnately lobed

pod: elongated capsule fruit consisting of two fused carpels and several seeds on a central wall

procumbent: spreading along the ground, but not as closely as prostrate

prostrate: growing flat to the ground

raceme: elongated inflorescence of individually stalked flowers, opening bottom to top

ray floret: minute flower with a petal-like projection (ligule), comprising part of a capitulum (flowerhead)

rhizome: a root stalk usually subterranean or growing just above the surface that produces shoot and root systems

rosette: ring of leaves surrounding the bottom of a main stem, usually on or near the ground

runner (also known as stolon): a spreading stem that runs along the surface of the ground and from which new shoots and plants can arise

schizocarp: fruit that, when ripe, divides along the septum (dividing wall) into two or more partial fruits

serrate: having a serrated/toothed edge

sessile: without a stalk

sori (singular sorus): clusters of sporangia (enclosures containing spores) attached to the underside of fern fronds

spikelet: individual flower of a grass, within a panicle

spore: single-celled reproductive units in non-flowering plants (ferns and bryophytes) designed for dispersal

spur: part of the flower – a hollow protuberance extending away from the inside of the flower

stellate: starry or star-shaped

stipule: leaf-like appendage arising below a leaf or main leaf stalk

suckering: spreading by means of suckers, i.e. new plants arising from roots

tripinnate: dividing, dividing again, and then dividing a third time (think fractals)

tussock: growth form in which many shoots of a plant grow close together (e.g. grasses)

umbel: inflorescence comprising several flowers or clusters of flowers, their stems arising from the same point (as in umbellifers)

vegetative reproduction: asexual reproduction of plants, in which a new, independent plant grows from individual plant parts

Photo credits

All photographs by the authors, except the following:

Pages 10, wall ferns – David Hawkins

Page 18, Sweet Alison – Krzysztof Ziarnek, Kenraiz, CC BY-SA 4.0

Page 21, Lesser Swine-cress (main image) – David Hawkins

Page 21, Lesser Swine-cress (inset) – Harry Rose, CC BY 2.0

Page 22, Danish Scurvygrass – Strobilomyces, CC BY-SA 3.0

Page 24, Rue-leaved Saxifrage – David Hawkins

Page 28, Navelwort – David Hawkins

Page 30, Fool's Parsley – Stefan Lefnaer, CC-BY-SA-4.0

Page 38, Common Mouse-ear – Dylan Peters, https://wildbristol.uk/

Page 42, Jersey Cudweed – Stefan Lefnaer, CC-BY-SA-4.0

Page 48, Eastern Rocket – Dylan Peters, https://wildbristol.uk/

Page 51, Welsh Poppy – Evelyn Simak, CC BY-SA 2.0

Page 71, Oxford Ragwort – Dylan Peters, https://wildbristol.uk/

Page 81, Yellow Corydalis – David Hawkins

Page 86, Common Mallow – Dylan Peters, https://wildbristol.uk/

Page 91, Red Valerian – Dylan Peters, https://wildbristol.uk/

Page 94, Henbit Dead-nettle – David Hawkins

Page 95, Common Fumitory (main and inset) – David Hawkins

Page 97, Wall Speedwell – AnRo0002, CC0

Page 98, Butterfly Bush – Dylan Peters, https://wildbristol.uk/

Page 100, Keeled-fruited Cornsalad – Dylan Peters, https://wildbristol.uk/

Page 106, Purple Toadflax – Dylan Peters, https://wildbristol.uk/

Page 111, Pellitory-of-the-Wall – David Hawkins

Page 125, Water Bent – David Hawkins

Page 129, Barren Brome – Dylan Peters, https://wildbristol.uk/

Page 130, Greater Quaking-grass (main image) – Matt Lavin, CC BY-SA 2.0

Page 130, Greater Quaking-grass (inset) – Gail Hampshire, CC BY 2.0

Page 132, Field Horsetail (main image) – Rob Hille, CC BY-SA 3.0

Page 132, Field Horsetail (inset) – David Hawkins

Page 133, Hart's-tongue – David Hawkins

Page 134, Rustyback – David Hawkins

Page 139, Common Liverwort – David Hawkins

Index

Published in 2025 by
Pelagic Publishing
20–22 Wenlock Road
London N1 7GU, UK

www.pelagicpublishing.com

A Field Guide to Urban Plants: Common Species of Pavements, Walls and Waste Ground

Translated by Iain Macmillan
Illustrations by Roland Spohn
Additional species accounts by David Hawkins

Originally published 2023 in German as *Das Wächst in Deiner Stadt*
by Franckh-Kosmos Verlags-GmbH & Co. KG
Pfizerstraße 5-7, 70184 Stuttgart

https://doi.org/10.53061/OTDC5460

A CIP for this book is available from the British Library

ISBN 978-1-78427-474-0 Pbk
ISBN 978-1-78427-475-7 ePub
ISBN 978-1-78427-476-4 PDF

EU Authorised Representative: Easy Access System
Europe – Mustamäe tee 50, 10621 Tallinn, Estonia,
gpsr.requests@easproject.com

Title pages: Common Chickweed *Stellaria media* with
Sheep's Sorrel *Rumex acetosella*
Facing contents page: Dark Mullein *Verbascum nigrum*
Cover image: Oxford Ragwort *Senecio squalidus*,
© James Common (@CommonByNature)
Back cover images by David Hawkins
Original design concept: Sigrid Walter, Würzburg
Typesetting: S4Carlisle Publishing Services, Chennai, India

10 9 8 7 6 5 4 3 2 1

Printed in the Czech Republic by Finidr